Strain Gage Based Transducers

Their Design and Construction

Prepared by
The Technical Staff of
Measurements Group, Inc.

Published by
Measurements Group, Inc.
Raleigh, North Carolina, 27611, USA

Library of Congress Cataloging-in-Publication Data

Strain gage based transducers.

Includes index.
1. Strain gages. 2. Transducers. I. Measurements Group. Technical Staff.
TA413.5.S85 1988 620.1'123'0287 88-12702
ISBN 0-9619057-0-0 (pbk.)

Foreword

The strain gage based transducer has been an important engineering tool since shortly after the invention of the strain gage in the late 1930's. During the past decade, however, strain gage transducer applications have grown at a remarkable rate, in both number and variety. In addition to their many industrial and engineering uses, these transducers are now appearing in business and consumer products with ever increasing frequency. As a result, there is currently a great deal of interest in transducer technology.

This monograph has been written to provide a systematic and comprehensive treatment of strain gage transducers — from the origin of such devices to present-day state-of-the-art practice. The emphasis here is on transducers which employ bonded metallic resistance strain gages as sensing elements; and the subject matter is divided into several sections, each of which focuses on an important aspect of the transducer as a measuring system. Although there are already many useful publications in this field, a more thorough treatment of transducer principles — particularly as affected by strain gage theory and practice — should prove helpful to both experienced practitioners and relative newcomers to the art.

It is assumed in this presentation that the reader is reasonably familiar with contemporary resistance strain gage technology. This implies a general understanding of gage operation, and the principal factors affecting gage performance, along with some awareness of the practical considerations involved in the installation and use of strain gages. On request, Measurements Group, Inc. will be pleased to furnish references for background study in strain gage fundamentals.

Table of Contents

1

Introduction

The resistance strain gage, because of its unique operational characteristics, has easily dominated the transducer field for the past twenty years or so.

Background

Broadly speaking, a transducer is a device which transforms one type of energy into another. A battery is therefore a transducer (chemical energy converted to electrical energy), as is an ordinary glass thermometer (heat energy converted into mechanical displacement of a liquid column).

We concern ourselves here with a specific class of transducers: devices which translate an input of mechanical energy into equivalent electrical signals for measuring and/or controlling the input phenomena. This type of electromechanical transducer is usually located at the source of the physical force or energy, and responds to its magnitude. The readout or control instrumentation can then be positioned at any convenient distance from the transducer, and connected to it by electrical wiring.

Common examples of such transducers are used to measure fluid pressure, weight, acceleration, displacement, torque, etc. The word "cell" is often used for convenience to describe a compact transducer (i.e., load cell, pressure cell). Certain forms of transducers have their own family names which usually derive from the physical phenomena they measure. Examples include accelerometers, extensometers and vibrometers.

Transducers can be, and are, manufactured on many different operating principles — resistive, inductive, capacitive, piezoelectric, etc. Miniature accelerometers for the measurement of high-range dynamic acceleration forces, for example, are usually constructed with piezoelectric sensing elements because of the resulting small size and weight, and the self-generated electrical output. Similarly, when some special aspect of the application requires it, capacitive or inductive sensors may be used. The bonded metallic resistance strain gage, however, because of its unique set of operational characteristics, has easily dominated the transducer field for the past twenty years or so.

The true origin of the strain gage transducer (and the strain gage itself, for that matter!) is lost in history. Lord Kelvin reported on the strain-induced resistance change of electrical wires in the 1800's. In 1908, Dr. St. Lindeck of Charlottenburg in Germany suggested what might be the first bonded strain gage pressure transducer. He was working with precision resistors which consisted of fine manganin wire wound on thin-wall brass tubes. The construction method involved coating the wire with shellac, and then baking the assembly to firmly bond the wire to the tube. In a study of factors affecting the stability of such resistors, St. Lindeck closed off the ends of the tubes, pressurized them internally to about 800 psi *(5.5 MN/m^2)*, and found that the coil changed resistance in a linear fashion with pressure. His report stated, *"A method of measuring high pressure can possibly be based hereon."* (emphasis by St. Lindeck). Figure 1.1 is an illustration from U.S. Patent No. 2,365,015, issued to E. E. Simmons in 1944, for a bonded strain gage pressure transducer. While it is certainly not true that "there is nothing new under the sun," it *is* difficult to be completely original!

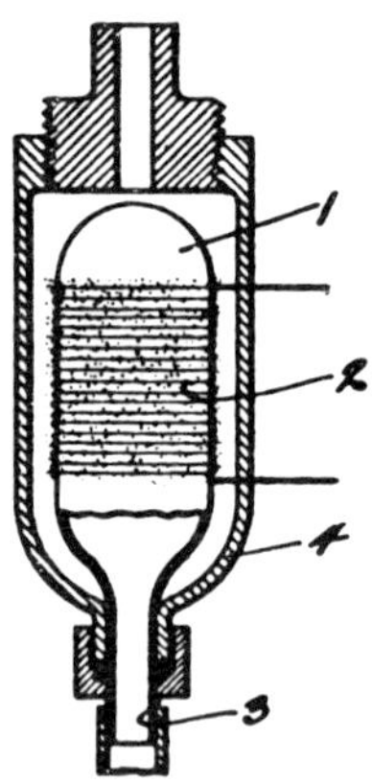

Fig. 1.1 — An early concept for a bonded strain gage pressure transducer.

E. E. Simmons and A. C. Ruge are generally credited with being co-inventors of the bonded wire strain gage. Ruge, in particular, was instrumental in introducing many practical forms of the bonded gage and various types of transducers which incorporate it. Unbonded wire strain gages were known and used in transducers, both before and after the introduction of the bonded gage, but they were much less versatile and less accurate.

In the 1950's, metallic foil bonded strain gages were introduced and quickly supplanted wire gages in most forms of strain gage transducers. Using foil instead of wire improved heat dissipation, reduced creep effects, and allowed much greater design freedom in adapting gage shapes and sizes to complex transducer geometries. Typical commercial transducer designs employing foil resistance strain gages are illustrated in Fig. 1.2.

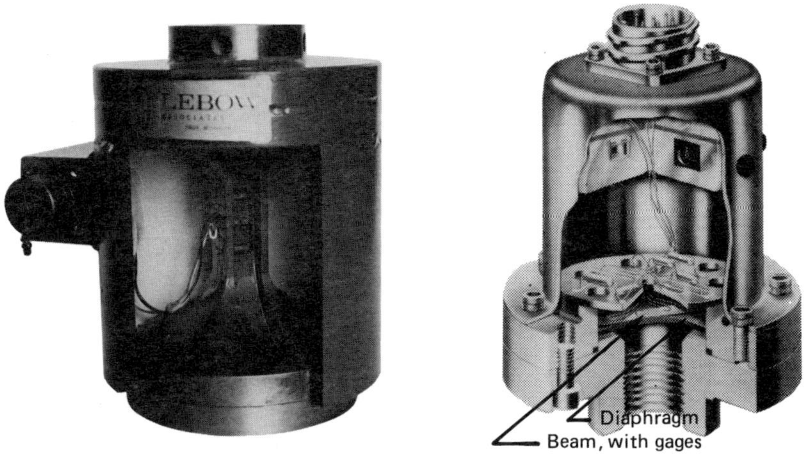

a. b.

Fig. 1.2 — Representative contemporary strain gage transducers: **a.** Cutaway load cell (courtesy of Eaton Corp./Lebow Products), **b.** Cutaway drawing of pressure cell (courtesy of Viatran Corp.).

Two other specialized developments in bonded resistance strain gages have had important impacts on the transducer field and are briefly described:

Semiconductor Gages

A spin-off from semiconductor technology, bonded semiconductor strain gages are produced in the form of gage patterns deposited on suitable insulating substrates, and as small bars or ribbons of doped silicon, equipped with fine leadwires and adapted for adhesive bonding to the transducer structure. High gage factors (20 to 100), small size, and high resistance characterize these piezoresistive gages. They are well suited for use on miniature, high-output transducers, although they are neither so practical nor so widely used as are conventional metallic gages in general-purpose, high-accuracy transducers. It is worth noting that semiconductor gages were originally considered advantageous because of their high output. This has less importance today because the same semiconductor technology which created the type of gage has also created smaller, less expensive amplifiers for use with conventional metallic gages.

Deposited Metallic Gages

Much work has been done in the last several years on the deposition of thin films of pure metals and metal alloys. Such metallic films can be deposited on pre-insulated transducer structures and used as bonded resistance strain gages. Deposition can be made through a suitable mask, resulting directly in the desired gage configuration; or the deposited film can subsequently be masked and etched into an appropriate gage shape. Potential advantages of such gages include elimination of organic adhesive layers (through the use of deposited inorganic insulating films) and low-cost, high-volume production methods. In actual practice, however, combining these advantages has not been easy. Furthermore, the performance of deposited films is significantly poorer in several respects than that of precision-rolled films of wrought alloys. Like semiconductor strain gages, therefore, deposited-film gages are primarily used in special-purpose transducers which can take proper advantage of their unusual, but limited, performance characteristics.

The Bonded Strain Gage as a Transducer Element

No single sensing principle is perfect for all transducer applications. And, like all other known sensors, the bonded strain gage has certain limitations. To take full advantage of its characteristics — to optimize a transducer design from cost and performance standpoints — it is essential to understand both the virtues and shortcomings of the strain gage, and the various forms they take.

First, the operating principles of the different sensor types should be well understood, and a transducer selected whose response mode is compatible with the measurement problem. Suppose one wants to measure displacement over a range of, say, one centimeter. In such a case, the most practical sensing element would normally be one which responds directly to displacement. Possible choices would include differential transformers (LVDT's), optical encoders, or linear potentiometers. The various specifications for size, accuracy, cost, operating environment and, to some extent, type of output signal desired, will dictate the final selection. In this example, a strain gage system would normally not be the first choice since the strain gage responds to *strain*, not displacement.

In some cases, the choice of sensor will depend more on special performance features needed in the transducer than on the primary actuation mode of the sensing device. If a displacement range of one millimeter were needed in the above displacement transducer, together with long life, wide frequency response, shock resistance, and a moderate restoring force in the actuator probe, a full-bridge strain gage network mounted on a cantilever spring might be a very appropriate solution. The point to be made is this: certain forms of transducers are well suited for bonded metallic strain gage sensors, and some are not. If strain gages are used where they are not a logical first or second choice, the end result will not be optimum.

The most significant operational characteristics of the bonded metallic resistance strain gage for transducer applications are tabulated below. Keep in mind that the gage responds to surface strain in a structure, and that in all normal uses a suitable elastic structure must be carefully designed into the transducer. This elastic element is often referred to as the "spring," since it usually provides the required resisting and restoring force. The configuration and material selections for the spring element are often the most critical decisions made in a successful transducer design. These topics are treated in detail in following chapters of the monograph.

Favorable Factors

1. ***Small size and very low mass:*** Important in minimizing inertial (shock) effects. Allows operation over a very wide frequency range — from d.c. (static) to over 50 kilohertz. The upper frequency limitation is set primarily by the strain gage grid size (gage length).
2. ***Fully bonded to basic spring structure:*** Together with (1) above, this freedom from critical mechanical connections — or "loose parts" — results in rugged, shock-resistant construction.
3. ***Excellent linearity over wide range of strains:*** In most cases, overall linearity will be limited by characteristics of the spring element rather than the strain gage.
4. ***Low and predictable thermal effects:*** Simple compensation and correction methods allow accurate operation over wide temperature ranges.
5. ***Highly stable with time:*** When suitably protected from overloads and destructive environments, strain gage transducers retain calibration very well over extended periods of time.
6. ***Relatively low in cost:*** Overall transducer cost is a complex and important aspect of the design process. Gage cost is seldom an important factor.
7. ***Circuit output is a resistance change:*** The resistance strain gage is a passive (as opposed to self-generating) sensor. Because it is essentially insensitive to supply-voltage frequency, it can be used with a.c. (carrier) or d.c. excitation systems. Gages normally operate in some form of bridge circuit, and these circuits can be easily arranged for altering transducer response through the addition or subtraction of moments and forces. This permits relatively straightforward design of complex transducers which respond to (or reject) multiple components of input force as required.

Limiting Factors

1. ***Thermal degradation:*** Gages employing organic materials as carriers, adhesives, and protective components are the most accurate, versatile, and easily used, but are limited in high temperature applications by the characteristics of the particular organics employed. Maximum operating temperatures for such transducers generally range from 150° F *(66° C)* to perhaps 500° F *(260° C)*, depending on specifications, time of exposure, and operating environment for the gages. Low temperatures do not cause serious problems, and transducers for use at cryogenic temperatures are easily designed.
2. ***Output signals are relatively low:*** Accurate, stable amplification and controlled excitation voltages are essential for accurate readout instruments. With modern electronics, however, this does not normally present a significant cost problem. System accuracy is often more limited by signal-to-noise ratio, which is quite good for strain gage transducers because of their low-reactance circuits and moderate ohmic resistances.
3. ***Careful installation procedures required:*** Bonded gage performance is strongly affected by the materials and procedures used to bond the gages in place, wire them into the circuit, and provide protection against handling damage and environmental factors. Without careful, rigorous control of every aspect of the installation process, predictable low-cost transducer production cannot be achieved.
4. ***Moisture effects:*** Simple, organic protective systems for bonded resistance strain gages can be effective, but will have only limited life when exposed to excessive moisture. With time, water will reach the strain gages and destroy or degrade their performance. For reliable long-term accuracy under such conditions, true hermetic sealing of the transducer is essential.

2
Load Cells

The most critical mechanical component in any load cell is generally the spring element.

Spring Elements

In this chapter, our attention is centered on force-measuring transducers, or "load cells," as they are frequently known. Although the strain gage load cell was developed and commercialized about 40 years ago, its principal area of application was, until recently, industrial — i.e., in process control, heavy machinery, test engineering, and the like. Over the past decade, however, the load cell has invaded (and largely conquered) the weighing field. With the need for scales having electrical output signals to operate the popular digital displays, and to interface with microprocessors and computers, the strain gage load cell has come to represent the most practical weighing means. Electronic scales based on strain gage load cells are now commonplace. One of the largest uses is in retailing, but others include postal and shipping scales, crane scales, laboratory scales, on-board weighing for trucks, agricultural applications, etc.

The most critical *mechanical* component in any load cell, or other strain gage transducer, is generally the *spring element*. Broadly stated, the function of the spring element is to serve as the reaction for the applied load; and, in doing so, to focus the effect of the load into an isolated, preferably uniform, strain field where strain gages can be placed for load measurement.

Implicit in this definition is the assumption that the strain level in the gaged area of the spring element responds in a linear-elastic manner to the applied load. In other words, the ideal transducer would be characterized by an unvarying, *proportional* relationship between the strain and the load. Achievement of this simple-sounding but elusive goal is central to all transducer design. The task is made difficult, to begin with, by the presence of numerous other operational and economic constraints which must be simultaneously satisfied to result in a practical, viable commercial transducer. Compounding the difficulty is the fact that a variety of effects, which would be of second- or third-order importance in normal engineering practice, can become highly significant in a precision instrument. Because of these considerations, we should thoroughly examine the design of spring elements before going on to other aspects of load cell design.

For purposes of this discussion, load cell spring elements will be divided into three classes, according to the type of strain field used. These are: (a) bending, (b) direct stress, and (c) shear, as schematically illustrated in Fig. 2.1. While different forms of all three types are used in contemporary transducer practice, the richest variety in configuration is undoubtedly found among the bending elements, and these will be described first. Preparatory to doing so, however, it will be fruitful to review a list of basic design considerations and criteria which apply to all transducer spring elements. With this background, it will be easier to appreciate the underlying reasons for some of the exotic configurations used in commercial load cells.

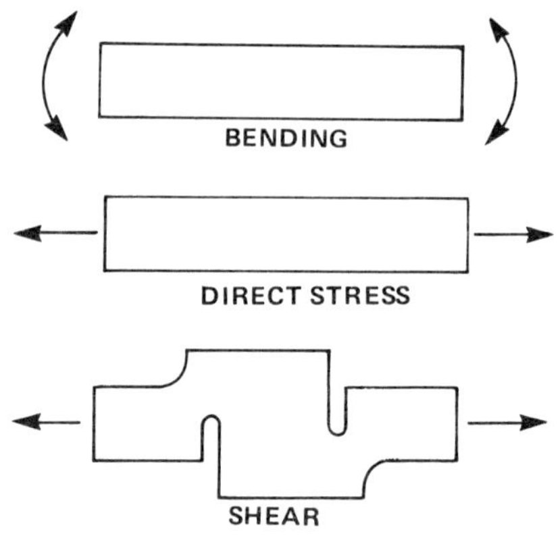

Fig. 2.1 — Three modes of spring element deformation commonly used in transducers.

Design Considerations

The following criteria are generally applicable to all load-cell spring elements. No interpretation should be made of the order in which they are presented, since the relative importance given to any of them depends upon the type and purpose of the particular transducer involved. Moreover, they are not independent of one another, but have numerous interactions. Good transducer design requires judicious, knowledgeable trade-offs and compromises among these criteria.

Natural frequency: Ordinarily, the natural frequency of the spring element should be as high as it can be made — consistent with the specified sensitivity and other operating requirements for the load cell. This normally calls for a rigid, low-compliance design, without unnecessary mass.

Appropriate strain level in the gage area at rated load: At an early stage in the spring element design it is necessary to establish the element proportions so that a predetermined strain level will be developed in the gage area when the unit is subjected to rated load. Based upon the combination of several different constraints (linearity of spring material response, fatigue life of strain gages, instrument compatibility, etc.), the level is frequently set in the range from 1000 to $1700\mu\epsilon$. With a four-gage fully active bridge circuit, $1500\mu\epsilon$ will produce a nominal output signal of 3mV/V of bridge excitation, based upon a gage factor of 2.0. The higher strain levels are used when it is necessary to achieve the same 3mV/V calibration in the presence of dissipative compensating circuitry.[1]

Uniform strain distribution in the strain gage area: Since the electrical output of the load cell is limited by the maximum allowable strain level in the gage region, this strain level should exist uniformly over the entire area of the gage grids to maximize the signal. When it can be accomplished, having the solder tabs of the gages lie on an area of lower strain will generally improve the fatigue life of the gages.

Lower strain levels throughout the remainder of the spring element: The strain magnitude in the gaged area of a spring element should not only be uniformly distributed and at the proper level to produce the desired full-load output signal, but normally it should also be the highest strain level anywhere in the entire spring element. In general, many of the load cell performance characteristics — e.g., fatigue life, linearity, freedom from creep and hysteresis — will be improved by minimizing strain levels throughout the structure. As Heraclitus expressed it some 2500 years ago, "everything flows." In our context, this implies that all solids, including the best of spring element materials when their behavior is observed closely enough, exhibit some degree of microplasticity under stress. Obviously, then, lower stresses mean smaller deviations from ideal linear-elastic behavior. They can also mean greater rigidity and higher natural frequency for the spring element.

Monolithic (one-piece) construction: Other factors being equal, repeatability, linearity, and freedom from hysteresis will be best in a load cell of which the spring element is machined entirely in one piece from a solid billet. Even though the deformation of a spring element under load may be very small, and not

---

[1]For example, temperature-sensitive resistors in the bridge supply to compensate for span shift with temperature. This and related topics are treated in Chapter 5.

visually discernible, it is nevertheless real and finite. Under these circumstances, nonintegral mechanical connections of any kind invite small movements and friction, and thus nonlinearity and hysteresis. Welded connections are also preferably avoided because of residual stresses and metallurgical effects which can limit fatigue life and promote microplastic behavior.

Design for ease of machining and gage installation: To be successful in the marketplace, a load cell or other transducer must be competitively priced. The process of accomplishing this within the framework of imposed functional constraints offers the opportunity for considerable ingenuity in element design. There is little to be gained by meeting all of the other design criteria in this list if the resulting spring element is unduly difficult and expensive to machine. It is equally important to fully appreciate the economics of strain gage installation. The cost of making a transducer-quality gage installation, even under ideal circumstances, is typically much greater than the cost of the gage itself. Therefore, the spring element should always be designed with the gage-installation process kept well in mind. Ordinarily (when feasible) a single, flat, external surface lends itself best to easy and rapid gage installation. If other constraints prevent this, then particular attention should be paid, during the design, to how the various installation procedures — surface preparation, adhesive application, gage placement, clamping, curing, wiring, etc. — will be performed.

Overload protection: It is the experience of transducer manufacturers that the most common single cause of transducer malfunction during field service is overloading. In some designs, the configuration of the spring element is suitable for the easy incorporation of a mechanical stop. In others, the case or enclosure of the transducer may be used to limit the deformation of the spring element. As a general rule, commercial load cells are designed to withstand overloads up to 150 to 200% of rated load without damage, and 300 to 500% overload before destruction of the load cell.

Selectivity of response: In most cases, the ideal load cell would respond to forces or force components along only one axis, while being insensitive to force inputs from other directions, and to applied couples as well. At the same time, in many instances, the response of the load cell should be independent of the point of load application, at least over some prescribed area. Although the ideal load cell behavior can never be achieved, it can be approached quite closely by sufficiently careful design of the unit. This frequently involves not only the spring element as a mechanical component, but also the placement of strain gages on the element, and the design of the supporting and/or enclosing structure.

Spring element deflection: In order that strains at a conveniently measurable level be induced in the spring element with applied load, it is necessary for the element to experience finite deformation. This means that the geometry of the spring element changes slightly under load, and that the point of load application is displaced correspondingly. Any change in geometry is inevitably accompanied by a degree of nonlinearity in response. Rigidity of the spring element is thus beneficial, not only in terms of natural frequency, but also in minimizing nonlinearities due to geometric changes. Furthermore, for any displacement of the point of load application, such displacement will have the least effect on the accuracy and linearity of the load cell if it is straight-line motion, along the load axis.

Thermal considerations: It might at first be assumed that the process of arriving at an optimum spring element configuration is purely a matter of good mechanical design, but such is not necessarily the case. Thermal effects must also be considered, and two in particular have a bearing on the spring element configuration. For one thing, a strain gage is a source of heat, which is largely dissipated into the spring element. To minimize destabilizing temperature effects on both the spring element and the gage response, the element configuration should be symmetrical with respect to the gage locations. Beyond this, a load cell in typical field service will often be subjected to a temperature gradient from one end to the other or one side to the other. The perturbing effects of such temperature gradients can be reduced by careful thermal design. For this purpose, the heat flow paths from the load cell enclosure to the spring element, and within the spring element, must both be considered. In particular, it is important to design for minimum thermal impedance between strain gages in adjacent arms of the same bridge circuit.

Bending Elements

Spring elements which employ beams or beam-like configurations subjected to bending moments are widely used in contemporary commercial transducers. There are several reasons for the popularity of bending elements, and these tend to arise from the intrinsic properties of beams. A beam is characteristically a high-strain, low-force structural member — compared, for instance, to a column of the same cross section. Furthermore, in the case of a beam with a cross section which is symmetrical about the bending axis, there are always two surfaces subjected to equal strains of opposite sign. This offers a convenient means for implementing a full-bridge circuit, by mounting opposed pairs of gages on the two surfaces. If the beam is reasonably thin, the arrangement will result in good temperature compensation because the temperature differences between gages can be kept very low. As a further advantage, beam configurations can often be designed to provide flat, open surfaces for ease of strain gage installation.

The Basic Cantilever Beam

The simplest (but not generally the best) beam configuration for a bending transducer is the basic cantilever beam, Fig. 2.2a. Pairs of longitudinally aligned strain gages are mounted on the upper and lower surfaces, near the root of the beam. Although low in manufacturing cost, and convenient for strain gage installation, this type of spring element embodies a number of shortcomings with respect to the preceding list of preferred design characteristics. As described in the following, some of the deficiencies can be overcome by design refinements, but others are fundamental to the cantilever configuration.

In a cantilever spring element, the only useful function served by most of the beam length is to convert the applied load into a bending moment in the gage area. At the same time, however, this portion of the beam length contributes significantly to the deflection at the point of load application, and to the mass subjected to displacement. As a result, the spring element tends to be low in natural frequency. The design can be improved somewhat by concentrating the deformation largely in the gage area, as shown in Fig. 2.2b. For the same strain level in the gage area, the deflection is now much lower, but the mass has been increased. Further improvement can be gained by decreasing the mass of the beam extension; and this can be accomplished, for example, by making the beam extension hollow as in Fig. 2.2c. The use of aluminum alloy instead of steel for the beam material will also result in lower deflection and higher natural frequency for the same strain level in the gage area.

With the foregoing design changes, the simple cantilever spring element has gotten noticeably less simple — and more expensive to fabricate. Unfortunately, it is still lacking in some of the desirable attributes of an optimum spring element. Because the displacement path at the point of load application is curvilinear, the point moves laterally as it deflects. Also, under these conditions, the load is no longer applied in a direction perpendicular to the beam axis. Unless these effects are eliminated or compensated for, both will tend to limit the ultimate accuracy and linearity of the transducer.

Another undesirable feature of the design as we now have it is that the strain distribution is not uniform over the grid length of the strain gage. The strain is highest toward the fixed end of the beam, and decreases linearly along the gage length. This effect is most prominent in short beams, since the strain gradient varies inversely with the beam length. We can overcome this deficiency by introducing still another design refinement — i.e., by varying the beam width in the gage area to form a constant-stress beam. The constant-stress condition can be obtained most easily by linearly tapering the sides of the beam as shown in the plan view, Fig. 2.2d. For this to achieve the desired result, incidentally, the projected extensions of the two tapered sides must meet at the point of load application.

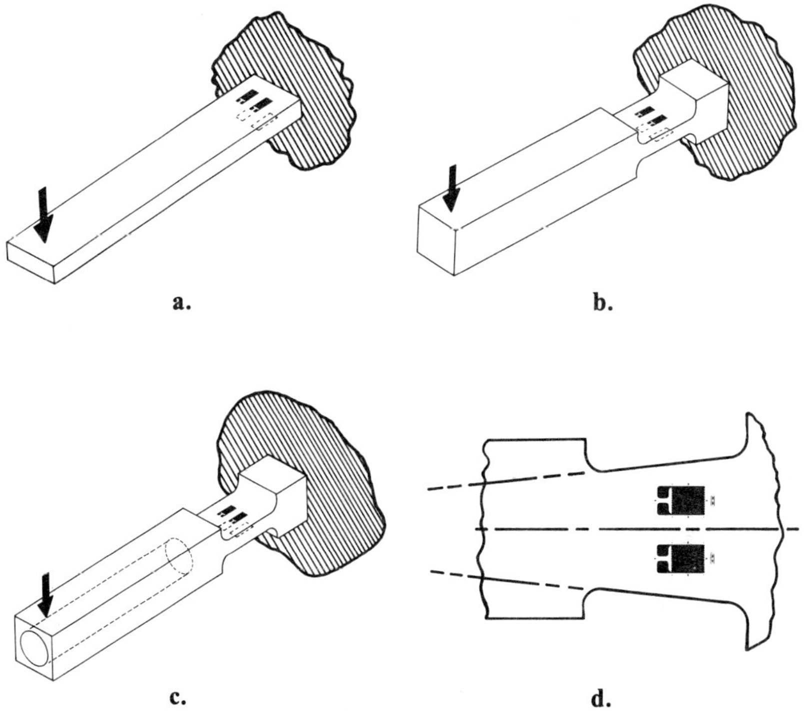

Fig. 2.2 — The basic cantilever-beam spring element, with successive refinements in configuration to overcome performance limitations (see text).

Performance Improvement

Both conceptually and historically, the cantilever beam represents the prototypical bending spring element for load cells. It should be evident from the foregoing, however, that we have at least approached the point of diminishing returns in our attempts to overcome the intrinsic performance limitations of the simple cantilever configuration. Because of this, load cell designers have explored a great variety of other bending-beam forms in their efforts to optimize transducer performance with respect to the previously listed design criteria. A few more-or-less typical configurations are described in the following section, and dozens more can be found among the load cell patents.

Accuracy specifications of 0.03% or less error are not uncommon in contemporary precision load cells — particularly those used in high-quality scales. The achievement of such accuracies requires that the load cell designer properly account for a number of second- and third-order effects which would normally be neglected in conventional mechanical design. For instance, small nonlinearities of 0.1% or less are completely ignored in the design of common mechanical and structural members. But, in transducer design, these and similar considerations make the difference between a commercially viable precision instrument and a makeshift sensor for relatively low-accuracy measurements. Because of the numerous simplifying assumptions incorporated in mechanics of materials solutions and, to a lesser degree, in the theory of elasticity, these engineering tools are, in themselves, inadequate for the optimum design of load-cell spring elements. They must be supplemented with analytical and experimental studies of the higher-order effects to achieve the refinements which distinguish first-quality transducers. And, of course, there is no substitute in this endeavor for good-old-fashioned experience and know-how.

Unfortunately, it is not possible to provide an algorithm, or a set of generalized rules, defining the design parameters for a universally optimum spring element. This is because the geometry of the spring element, and its modes of loading and deformation, are affected by too many constraints — rated capacity, measuring range, physical size, natural frequency, etc. — which appear in different combinations for different applications. The best that we can do in the following descriptions of bending spring elements is to note some of the more prominent higher-order effects. The practical significance of the effects for a particular design will obviously depend upon both the specifications for the design and the geometric details of the spring element. It should be noted that most of the spring element configurations shown here have been the subjects of one or more patents. While some of the basic patents date back to the 1950's or earlier, and have long-since expired, others may still be in force. Because of this, readers who may be contemplating the design or manufacture of a load cell would be well advised to consult the patent literature before doing so.

Multiple-Bending Elements

For certain types of applications, the characteristics of the straight cantilever beam can be improved upon by designs which induce "multiple bending" (reversed curvature) in the beam element. Consider, for instance, a beam which is built-in at both ends, and loaded at the center, as shown in Fig. 2.3. The figure also includes the bending moment and deflection diagrams, with the deflection *greatly* exaggerated for clarity. (Strain gage locations are indicated in this and subsequent sketches by short line segments labeled *T* and *C* for tension and compression, respectively.) Although this exact configuration is not widely used in commercial load cells, the structural simplification is convenient for illustrative purposes. The potential advantages of the design include intrinsic stiffness and straight-line motion of the point of load application as the beam deflects. The spring element also lends itself to relatively easy installation of a full-bridge strain gage circuit on the upper surface of the beam. Some degree of nonlinearity in output can be expected, however, because of the membrane stress produced in the beam (as it deflects) by the rigidly spaced end supports. Additionally, as for most flexural spring elements, it is necessary to vary the section modulus of the beam along its length if the strain gages are to lie in nearly uniform strain fields.

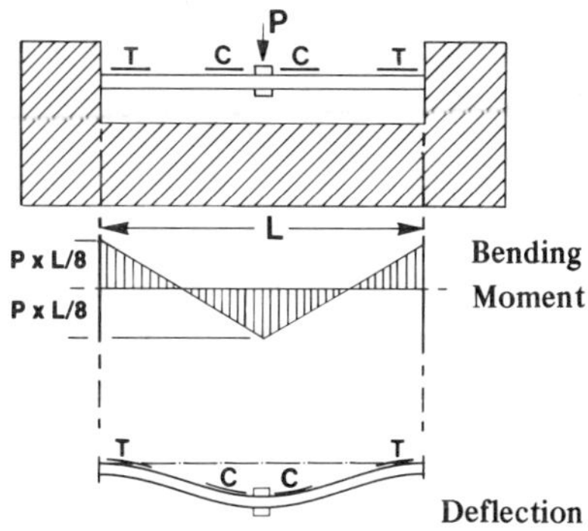

Fig. 2.3 — Generic beam arrangement for inducing reversed bending.

An alternative configuration, closer to some commercial designs, is shown in Fig. 2.4. This spring element has generally the same bending moment distribution and deflection pattern as that in Fig. 2.3 and retains essentially the same advantages except that the compliance is twice as great if the dimensions are otherwise the same. Because the end restraints are free to move laterally as the upper and lower beams deflect, the membrane stress is eliminated. Any such motion, however, represents a small change in the moment arm of the applied load, which can manifest itself in the form of nonlinear response if the ratio of the deflection to the beam length is great enough.

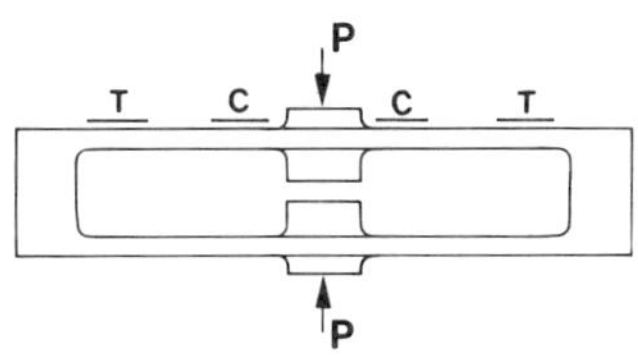

Fig. 2.4 — Reversed-bending beam configuration with membrane stresses eliminated.

Figure 2.5 illustrates a very simple means for producing effectively the same mode of double bending that occurs in either half of the sensing beams in Figs. 2.3 and 2.4. In this case, pairs of strain gages are mounted side-by-side on one surface of the beam, or back-to-back on opposite surfaces, to implement a full-bridge circuit. In the form shown here, the design is sensitive to both the location and direction of the applied load. To function properly, the design must incorporate features to assure that loading can occur only along the intended axis.

A significantly improved form of the preceding design is shown in Fig. 2.6, where the load sensing is accomplished with two beams, joined by relatively massive sections at both ends. With this configuration, externally applied couples are counteracted by axial forces in the sensing beams, minimizing the effects of off-axis loads. One of the drawbacks of the design, in the proportions shown, is its excessive compliance. The deflection which takes place in the beam segments between gage locations not only increases the compliance of the unit, but also degrades the linearity. Better load cell performance can be obtained by either shortening the beams or increasing the beam thickness between gage sites. *Such design changes should be made, of course, with full consideration of the shear loads which must be borne by the element.* Figure 2.7 illustrates a commercial transducer design which represents a simple, elegant, and eminently practical solution to the problem. Although strain gage installation and inspection are more difficult when gages are located inside of a hole as indicated in the figure, such arrangements are particularly well suited to protecting the gages and sealing them from the environment.

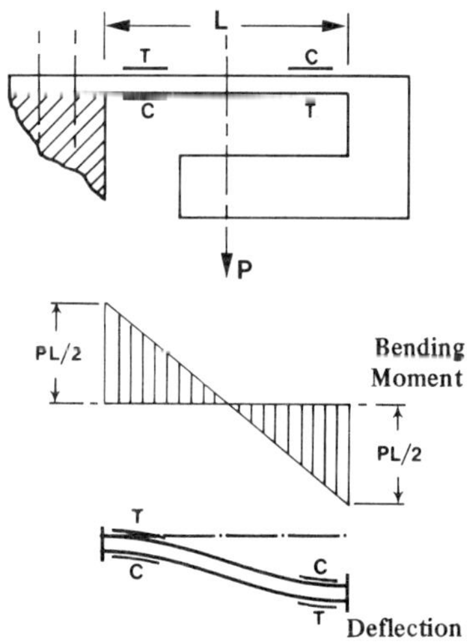

Fig. 2.5 — Reversed bending induced by folded cantilever beam.

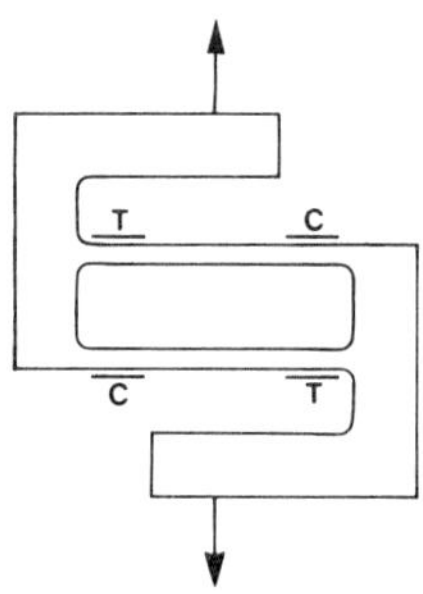

Fig. 2.6

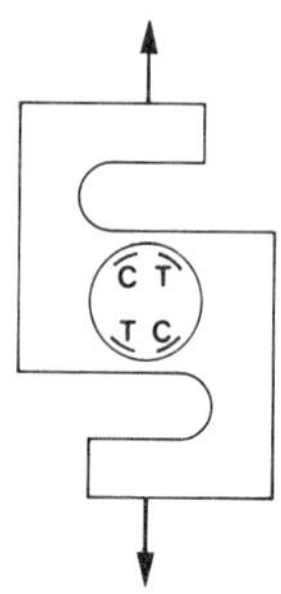

Fig. 2.7

Coupled dual-beam spring elements:
Fig. 2.6 — Basic configuration;
Fig. 2.7 — Compact, low-compliance design.

Various forms of the coupled dual-beam arrangement are widely used in load cells for weighing applications. In this case, it is especially advantageous to employ a spring-element configuration characterized by an output which is independent of the point of load placement. For small scales, the pan of the scale can be mounted directly on the spring element; and, in well-designed systems, the indicated weight will be the same, regardless of where the weighed object is placed on the pan. In Fig. 2.8, for instance, if analyzed at the level of first-order effects, the output of the gages, when combined by the Wheatstone bridge circuit, is the same whether the load is placed on the scale pan at point *A*, or *B*, or *C*.

The extraneous couples caused by off-center loads in Fig. 2.8 are counterbalanced by a couple composed of equal and opposite axial forces in the upper and lower beams. Although these forces cause corresponding strains in the beams, they do not directly change the output because they are cancelled in the bridge circuit. It is worth noting, however, that for any finite beam deflection, the axial forces affect the moment distributions in the beams, producing distinctly nonlinear behavior. Fortunately, the nonlinearity in the upper beam is opposite and complementary to that in the lower beam. When the strain gages are located as in Fig. 2.8, the two nonlinearities are also cancelled in the bridge circuit. As in all spring elements of this class, there still remains the small nonlinearity due to the decreasing primary moment arm with deflection of the beams.

In general, for any bending spring element, linearity will be enhanced by minimizing the ratio of the deflection (at rated load) to the length of the sensing beam, thus minimizing the change in shape of the element. For scale applications, however, the absolute deflection at rated load is also very important because (unlike the relative deflection) it directly determines the compliance of the element. The spring element and the object being weighed form a mass/spring system, the natural frequency of which varies inversely as the square root of the compliance. The design sketched in Fig. 2.8 is unnecessarily compliant. It can be markedly improved, with no loss in sensitivity, by thickening the beams everywhere except at the gage locations. A rather simple and popular way of accomplishing this in commercial load cells is exemplified by the configuration shown in Fig. 2.9, often referred to as the "binocular" design.

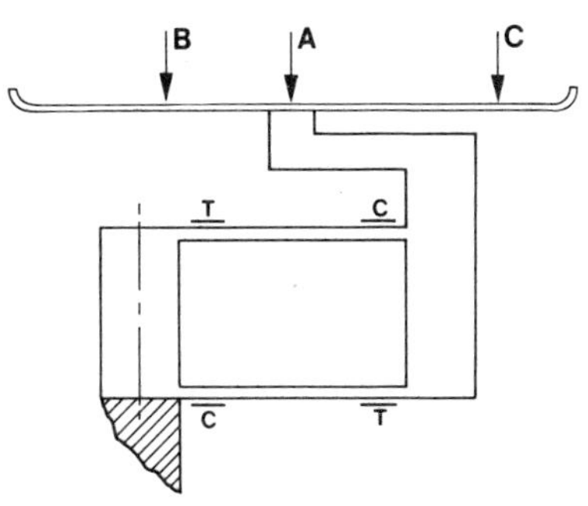

Fig. 2.8 — Dual-beam spring element for reduced sensitivity to off-axis loads.

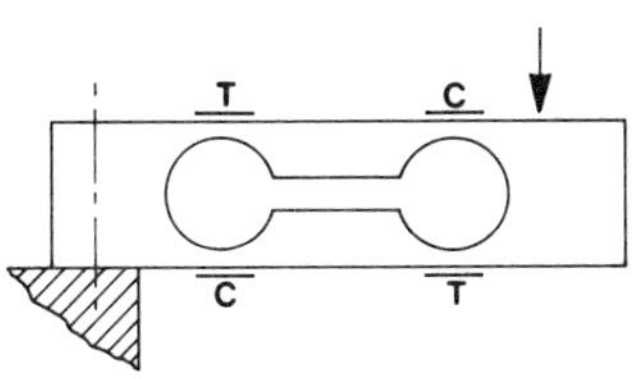

Fig. 2.9 — Popular "binocular" spring element, widely used in low-capacity scale applications.

The coupled dual-beam arrangement is recognizably present in many other load cell configurations, such as that in Fig. 2.10. This design has the same characteristic as the preceding dual-beam designs in that the bending moments in the sensing beams are independent of the point of load application. It also provides a feature that is intrinsic to the designs in Figs. 2.5 and 2.6; namely, elimination of the axial forces in the beams. If the load is applied to the right of center in the figure, the upper beam is in tension and the lower one in compression. If the load is applied to the left of center, the signs of the forces are reversed. A centrally applied load produces no axial forces in the beam whatsoever.

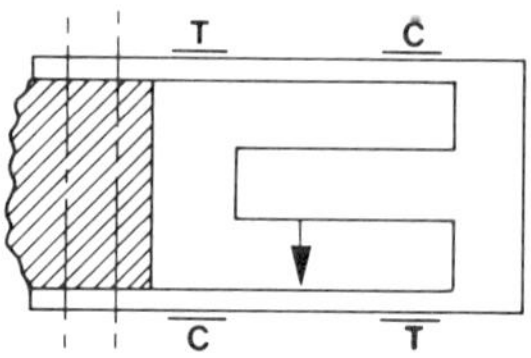

Fig. 2.10 — Centrally loaded dual-beam arrangement to eliminate direct stress in sensing beams.

Still another configuration is sketched in Fig. 2.11. In this case, the upper and lower members serve both as flexures and reactions for externally applied couples. Bending moments due to the vertical component of load are sensed in the central measuring beam. The flexures are locally thinned to minimize bending resistance while retaining sufficient cross section to carry the axial forces.

The spring elements discussed above have all been sketched as "two-dimensional" or plane structures, and sometimes in greatly simplified form to illustrate the mechanical principles involved. Spring elements which are optimized for linearity, compliance, maximum uniform strain field at gage locations, etc. often have numerous geometric refinements not shown here. The better designs are usually machined from solid billets of metal as integral structures. These will also have generous fillets at the junctures of connected sections, and may incorporate other features such as tapered sensing beams to maximize the strain gage output.

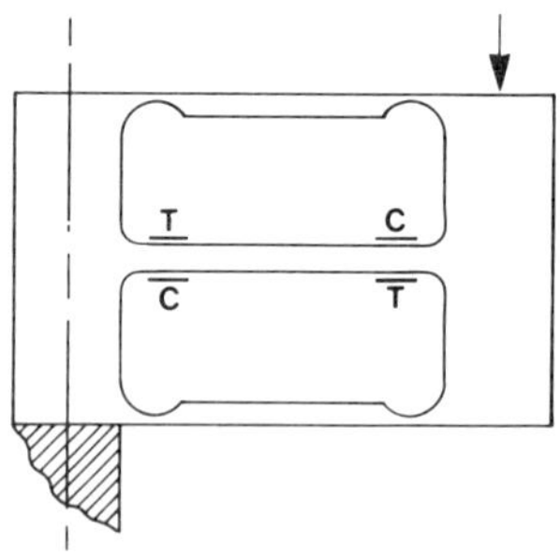

Fig. 2.11 — Triple-beam design for isolation of sensing beam from externally applied couples.

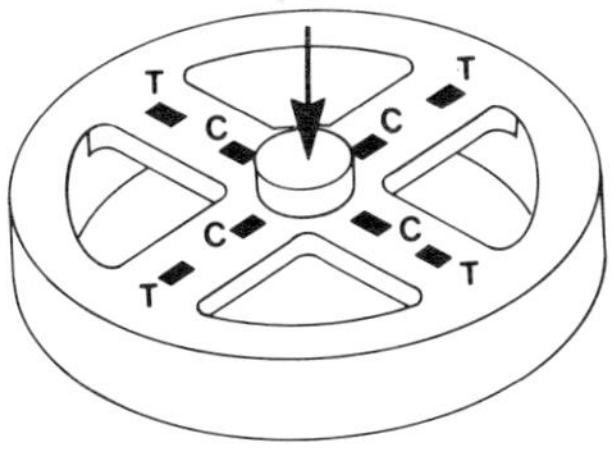

Fig. 2.12 — Wheel-shaped spring element, adaptable to low-profile transducers.

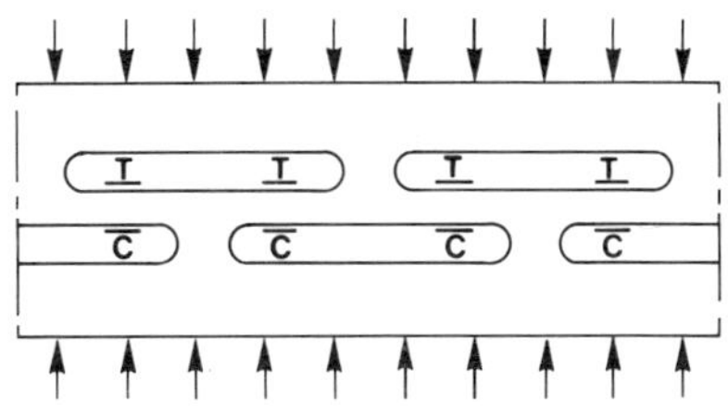

Fig. 2.13 — Developed view of radially slotted, axially loaded annular spring element.

When a "low-profile" spring element is required, the principles employed in the foregoing planar designs are sometimes adapted to round configurations. Fig. 2.12, for example, illustrates a wheel-like spring element in which the "spokes" serve as sensing beams. This design is operationally similar to the configuration shown in Fig. 2.3. Variations on this theme are numerous. Shown in Fig. 2.13, for instance, is a developed partial view of a slotting arrangement used to form beams in double bending around the periphery of an annular spring element.

Rings

Another type of bending spring element, which ranks with the beam designs in terms of the number and variety of its implementations, is the ring. The ring-shaped spring element also has a long, classical history in measurement technology, stemming from the well known Morehouse proving ring, once universally used to calibrate materials testing machines. Although ring-type spring elements always involve bending, direct stress is also intrinsic to the configuration, and the combination of the resulting two deformation modes provides the primary distinction from pure beams.

A basic ring design is shown in Fig. 2.14. The strain distribution in the ring is a complex function of the geometry, and is significantly affected by the design details of the bosses, but the distribution illustrated in the figure is reasonably representative for constant-thickness rings. Because the bending moment does not vary significantly in the region of the horizontal diameter, the strain distribution is nearly uniform in this area.

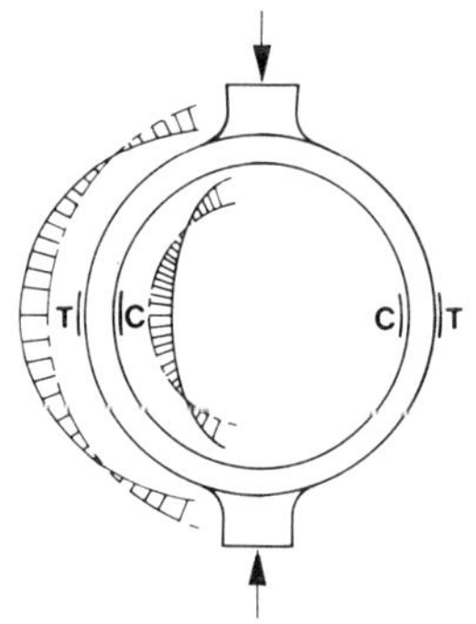

Fig. 2.14 — Simple ring configuration, based on classical Morehouse proving ring principle.

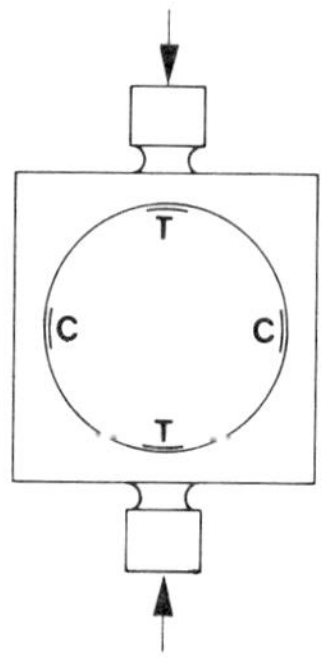

Fig. 2.15 — Improved ring design for low compliance, less sensitivity to off-axis loads, and reduced manufacturing cost.

The ring design in Fig. 2.15 offers several advantages over the classical form. To begin with, it is obviously easier, and much less costly, to fabricate. The squared ring also decreases the compliance of the spring element, and correspondingly improves the linearity. At the same time, the flexural stiffness at the junctures of the bosses and the ring has been reduced to minimize the sensitivity of the element to off-axis load components. There are countless other designs based on the presence of a stress-concentrating hole and / or lateral notches in an axially loaded member. A representative configuration, taken from a 1964 U.S. patent (No. 3,315,203), is sketched in Fig. 2.16.

In adapting the ring concept to different load cell specifications for capacity, physical size, etc., the designs sometimes deviate so far from a conventional ring in appearance that their classification as such becomes arguable. The configuration in Fig. 2.17, for instance, is placed here with the rings only because the gaged area of the spring element behaves in a ring-like fashion. The load-sensing strain measurements are made on sections subjected to combined direct and bending stress, with the bending moment uniformly distributed in the gaged area. Although similar in appearance to the binocular design of Fig. 2.10, the mode of loading and the strain gage locations give it quite different operating characteristics.

The bending spring elements pictured in this section represent only a small fraction of the many styles and configurations which have appeared in commercial load cells and in the patent literature. The great variety in design arises primarily from attempts to achieve the desired accuracy while meeting different

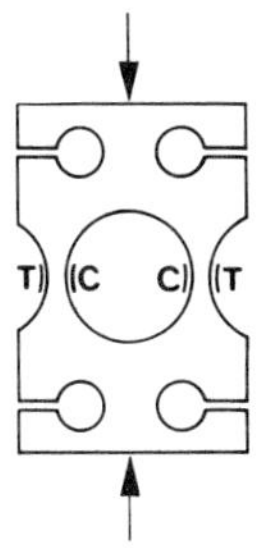

Fig. 2.16 — Complex spring-element geometry derived from ring concept.

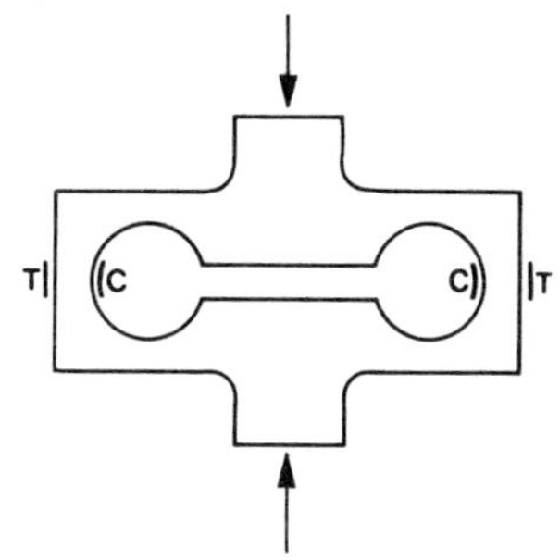

Fig. 2.17 — Variation of ring design, similar in appearance to "binocular" spring element, but very different in operating characteristics.

functional and operating requirements such as full-load capacity, sensitivity, physical size, manufacturing cost, etc. Although the discussion here has been limited to a rather broad look at various mechanical arrangements, following sections more closely examine some of the details which critically affect the interactions between the strain gages and the spring elements. It is often these latter considerations which strongly influence important load-cell performance such as sensitivity, stability, and repeatability.

High-Capacity Load Cells

The evolution of beam-type load cells has been traced from the basic cantilever beam, through a number of refinements, to a variety of more sophisticated forms with generally superior properties. Multiple-beam spring elements such as those described in the preceding section are currently very popular, and can be found in many commercial transducers — particularly in low-capacity units. It is the need for this last qualifier which leads to the subject matter of the present section.

Although the multiple-beam designs have good overall characteristics, including linearity and insensitivity to point of load application, they do not lend themselves well to being scaled up for higher load cell capacities. As the capacity of the load cell rises, so does the size of the spring element, along with its mass and, usually, its deflection at rated load. Because of these considerations, spring elements based on the measurement of bending strains are not commonly used in load cells with capacities greater than about 1000 lb *(~4 kN)*. Instead, transducer designers ordinarily turn to one of two other configurations — the shear web or the column — to achieve very high capacities in compact, low-compliance spring elements. Of these, the shear web will be discussed first for the sake of continuity, since the mode of loading for this type of spring element is often bending.

Shear-Web Load Cells

The principle of the shear-web spring element is illustrated in Fig. 2.18. In this case, it takes the form of a cantilever beam which has been designed with a generous cross section (with respect to the rated load) to minimize the deflection. Under these conditions, the surface strain along the top and bottom of the beam would be too low to produce the 2mV/V output normally expected of a strain gage transducer. At section *A-A* of the beam, however, a recess has been machined in each side, leaving a relatively thin web in the center. Just as in a structural I-beam, most of the shear force imposed by the load is carried by the web, while the bending moment is resisted primarily by the flanges.

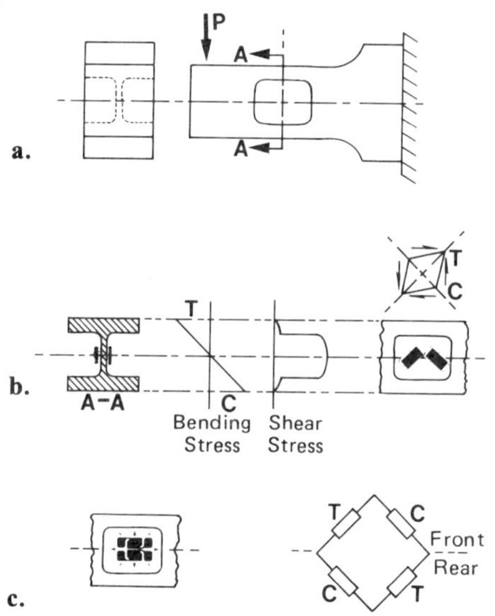

Fig. 2.18 — Principle of the shear-web load-cell, with stress distribution and circuit arrangement.

The cross section *A-A* is shown in Fig. 2.18b, along with the shear and bending stress distributions on this section. At the neutral axis, where the bending stress is zero, the state of stress on the web is one of pure shear, acting in the vertical and horizontal directions. As a result, the principal axes there are at ±45 deg to the longitudinal axis of the beam, and the corresponding principal strains are of equal magnitude and opposite sign. Pairs of strain gages, with their gridlines oriented along the principal axes, are installed on both sides of the web and connected in a full-bridge circuit for load measurement.

One of the advantages of the shear-web spring element is its slow sensitivity to variations in the point of load application. Static equilibrium considerations decree that the vertical shear force on every section of the beam to the right of the load (in Fig. 2.18a) be the same, and exactly equal to the applied load. Thus, the shear in the web should be independent of the point of load application (along the beam centerline), as long as the load is applied to the left of the web. If the strain gages sensed only the shear-induced strains, the bridge output would be unaffected by the position of the load or by other bending moments in the vertical plane.

Since the gage grids are necessarily finite in length, however, and thus span a small distance above and below the neutral axis as indicated in Fig. 2.18b, their outputs are also slightly affected by the bending strains in the web. With the grids centered on the neutral axis, the tensile and compressive bending strains above and below the axis tend to be self-cancelling in each grid. But the cancellation is usually less than perfect because of small asymmetries in the spring element and strain gage installation.

An alternative gage arrangement is shown in Fig. 2.18c. If the gridline directions on the back face of the web are made perpendicular to those on the front face, and if the gages are connected in the bridge circuit as indicated, bending effects are cancelled in the bridge. With this arrangement, the bridge output is theoretically independent of the bending moment, and thus of the load placement. The same arrangement serves to cancel any bending strains which may occur due to side loads on the beam.

Because of higher-order effects tending to couple the shear and bending strains, it is always preferable to design the beam for the lowest practicable bending moment in the shear web. This would seem to suggest the use of very short beams; but the point of load application must be far enough away from the shear web (á la St. Venant's principle) so that the web behavior approximates the ideal described here. One way to minimize the bending moment in the shear web is indicated schematically in Fig. 2.19. In this case, the bending moment at the center of the web is zero; and, for a given beam length and rated capacity, the bending moment throughout the beam is halved.

Shear-web spring elements are not limited, of course, to cantilever beam configurations; and a variety of other designs can be found in commercial load cells. Figure 2.20, for example, illustrates what is effectively a simply supported beam, because of the flexures at both ends. In this design (for which, we understand, patents are pending), four gages, one on each side of each web, make up the full-bridge circuit. The gage placement and circuit arrangement provide cancellation of bending strains due to either off-axis loads or side loads.

Another type of shear-web spring element is shown in Fig. 2.21. In this and similar designs, the element consists of a metal block in which holes or slots have been machined to form webs subjected to direct shear under axial load. A further and final example is given in Fig. 2.22 where the shear webs are produced by drilling a hole longitudinally through the beam. Strain gages oriented at ±45 deg to the beam axis are then installed inside of the hole to sense the shear force (see U.S. Patent No. 4,283,941).

In addition to their characteristically low sensitivities to misaligned load vectors, shear-web spring elements offer several other advantages to the load cell

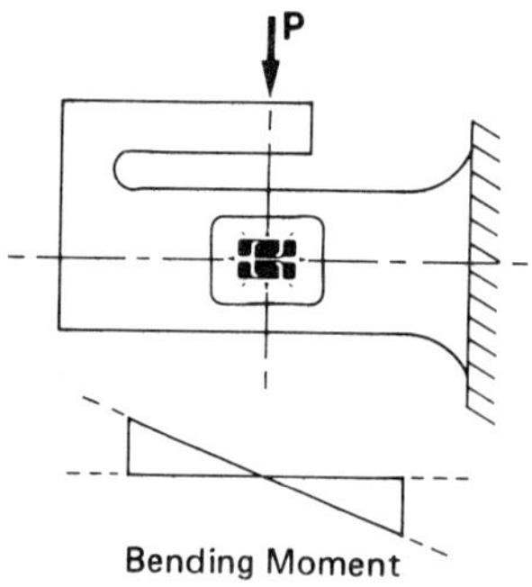

Fig. 2.19 — Shear-web spring element with zero bending moment at the gaged section.

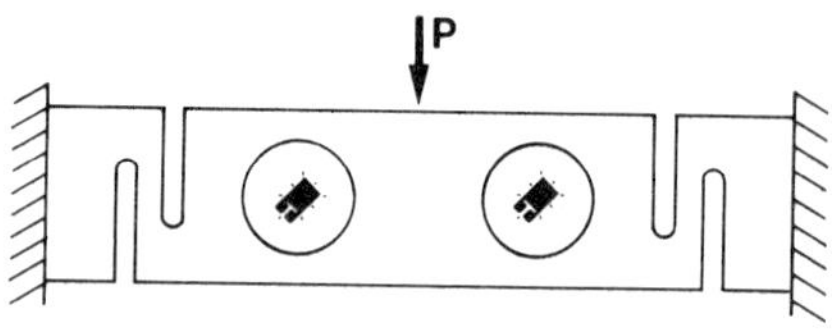

Fig. 2.20 — Dual shear webs in a beam configuration designed to cancel the effects of the off-axis and side loads.

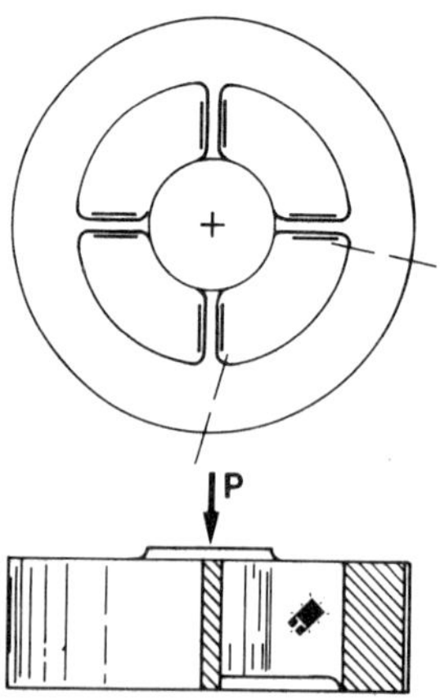

Fig. 2.21 — Spring element in wheel form, with radial webs subject to direct shear.

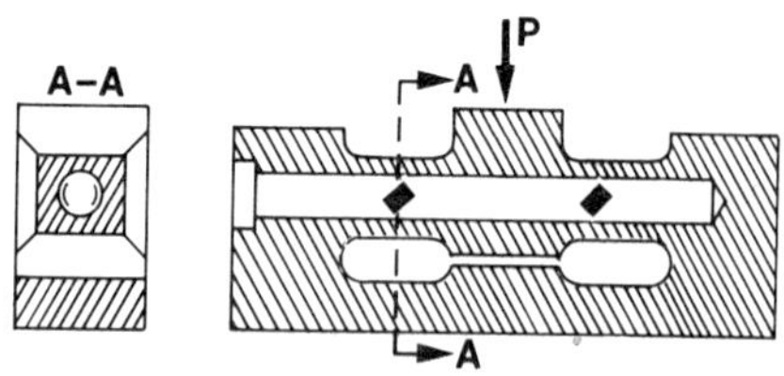

Fig. 2.22 — Shear webs formed by drilling a longitudinal hole in the load-sensing beam.

designer. Usually, for instance, they can be made very low in overall height for a given load capacity. Furthermore, the strain gages are commonly located in some form of recess where, even though more difficult to install, they can readily be sealed and protected from environmental effects. These features, along with low compliance and good linearity undoubtedly account for the growing popularity of shear-web spring elements for higher capacity load cells. They are not, however, well adapted to very-low-capacity applications. This limitation occurs because a web which is thin enough to produce the required output strain under rated load is apt to be too thin for elastic stability.

Column Load Cells

The column load cell has a history which dates back to the earliest strain gage transducers. As referred to here, the column spring element consists of one or more cylindrical members of the general form shown in Fig. 2.23 (but not necessarily circular in cross section). The spring element is intended for axial loading; and typically has a minimum of four strain gages, two in the longitudinal direction, and two oriented transversely to sense the Poisson strain. The gages are connected as indicated to form a full-bridge circuit — but not a fully *active* bridge circuit because the transverse Poisson strain is much smaller than the axial strain. When designed with the appropriate supporting structure and means for introducing the load into the column, this type of load cell can be made either for compression or for tension service, or for both.

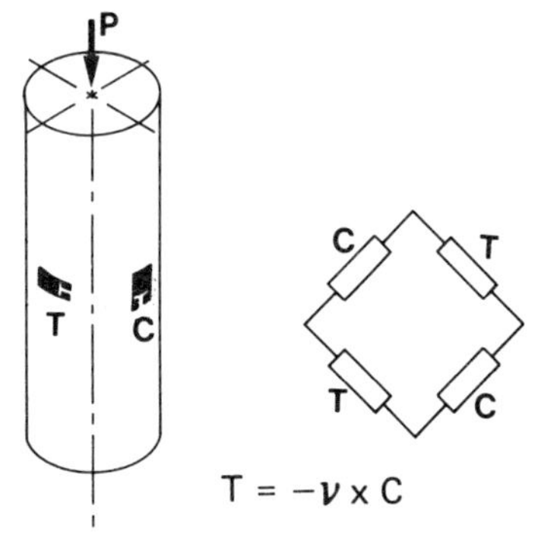

Fig. 2.23 — Principle of the column load cell.

Although conceptually simple, the column spring element embodies a number of special characteristics which require consideration when attempting to achieve a high-accuracy load cell. For example, the column is preferably made long enough with respect to its cross section so that a uniform strain field, unaffected by end conditions, is established at the strain gage locations. This normally requires that the cylindrical portion of the column be, say, five times as long as the largest cross-sectional dimension. Since the column configuration is subject to second-order effects from off-axis load components, provision must be made for isolating the axial load output from these. A representative design for a practical column load cell is shown in the simplified cutaway sketch of Fig. 2.24. The two diaphragms at the upper end of the column serve to minimize transmission of extraneous vectors and couples to the column.

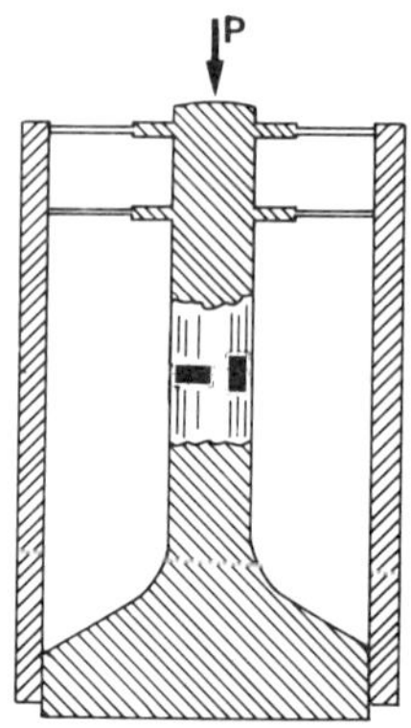

Fig. 2.24 — Representative design for a practical column load cell.

Deservedly or otherwise, column load cells have a reputation for inherent nonlinearity. The deviation from linear behavior is commonly ascribed to the change in cross-sectional area of the column (due to Poisson's ratio) which occurs with deformation under load. Because of this effect, the stiffness of the spring element continually increases as it is loaded in compression, and decreases with tension loading. Implicit in the latter statement is the assumption that the elastic modulus remains constant and independent of the accompanying density changes. More likely, the modulus tends to increase slightly in compression and decrease in tension, aggravating the area effect. Even though these modulus variations are too small for detection in ordinary materials properties tests, they can still be significant at the levels of precision involved in modern strain gage transducers. Pending the availability of more accurate data on elastic modulus variation with stress, we can at least estimate the nonlinearity due to the area change. Assuming a circular cross-section column, it can readily be calculated that the nonlinearity caused by the area change is approximately 0.003% per $100\mu\epsilon$ of longitudinal strain in the column.

Actually, the area and modulus effects described above are only two of several which seem to arise, directly and indirectly, from the nonsymmetric stress distribution in a column. With both bending and shear-web spring elements, in contrast, there are usually equal volumes of metal subjected to numerically equal tensile and compressive stresses, thus providing numerous mechanisms for internal self-compensation of nonlinearities — but the same is not true for column spring elements.

Consider, for example, the column load cell nonlinearity arising from the fact that all four arms of the bridge circuit are not equally active. As shown in Measurements Group Tech Note TN-507, *Errors Due To Wheatstone Bridge Nonlinearity,* this error occurs whenever the resistance change in one arm of the bridge circuit is not matched by an equal and opposite change in the adjacent arm. Because the resistance changes in the transverse strain gages on a column are only a third or less of those in the longitudinal gages, the bridge output is always nonlinear. In magnitude, the error is about 0.007% per 100$\mu\epsilon$ of column axial strain. The sign of the error is such that it is always opposite to the area-change nonlinearity evaluated above. Thus, the Wheatstone bridge nonlinearity not only cancels, but overcompensates for the geometric nonlinearity.

Described in U.S. Patent No. 3,034,346 is a very interesting technique for incorporating nonlinearity compensation in virtually any type of load cell. The method involves installing an additional strain sensor on the spring element, and connecting it in series with the bridge power supply rather than in the bridge itself. The output of the gage thus serves as a feedback for adjusting the bridge voltage in the opposite direction to that of the nonlinearity error. Depending on the material of the gage grid, and on the position and orientation of the gage on the spring element, approximate compensation can be achieved for almost any monotonically varying nonlinearity error.

Column spring elements take on a wide variety of forms in designers' attempts to optimize the load cell in terms of both production and performance considerations. The column cross section may be square, for example, instead of circular; or it may be circular with flats machined on four sides to facilitate strain gage installation.

Although column load cells are usually quite compact in transverse directions, they can become inconveniently tall and bulky when designed for very high loads. One means of obtaining a high-capacity, low-profile column load cell is shown in Fig. 2.25. In this type of design, the load is carried by four or more small columns, each with its own complement of strain gages. The corresponding gages from all of the columns are connected in series in the appropriate bridge arms. The result is not only a very compact configuration, but one which provides improved discrimination against the effects of off-axis load components.

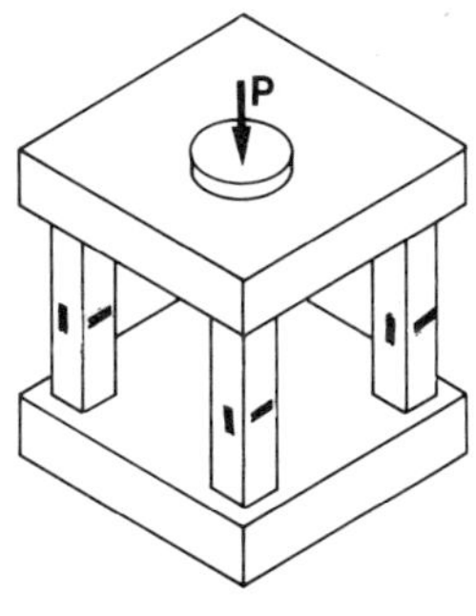

Fig. 2.25 — Four-column load cell for increased capacity.

The column spring element can also be made hollow as in Fig. 2.26. For the same net cross-sectional area (and therefore the same compliance with respect to purely axial loads), the hollow cylinder is more resistant to bending because of its greater area moment of inertia. This concept can also be implemented in a

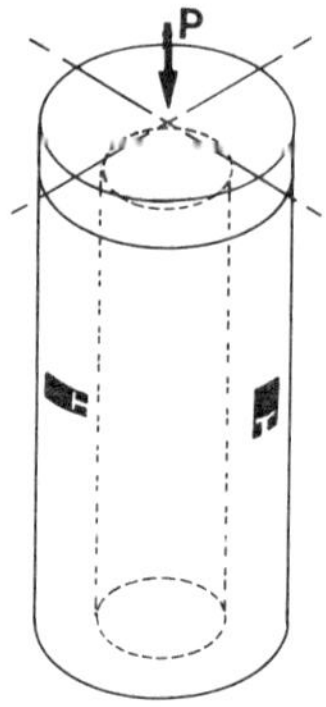

Fig. 2.26 — Hollow cylindrical load cell is less sensitive to bending moments.

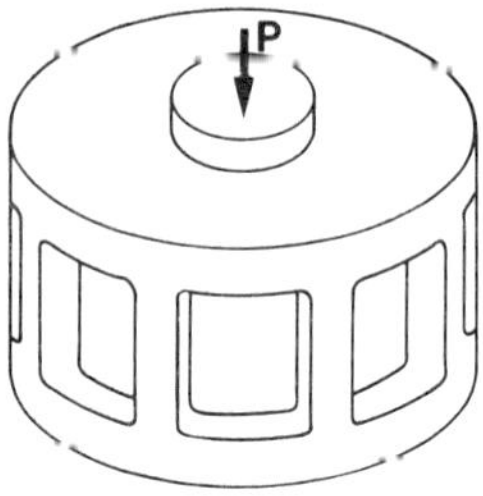

Fig. 2.27 — Slotted-cylinder, multiple-column configuration for high-capacity, low-profile applications.

low-profile, multiple-column configuration such as that shown in Fig. 2.27. Still another method of increasing the area moment of inertia for a given cross-sectional area is illustrated in Fig. 2.28. The H-section column has the important advantage of allowing the strain gages to be positioned much closer to the neutral axis of the column. In this location they are correspondingly less sensitive to any bending strains which may occur in the spring element.

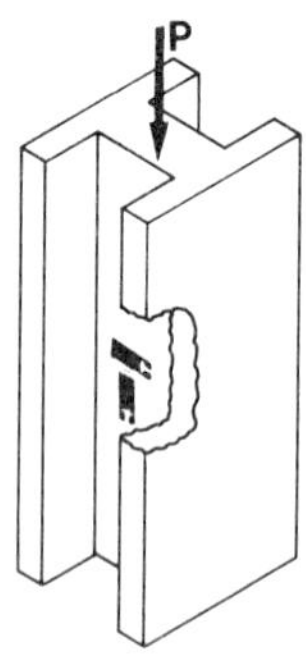

Fig. 2.28 — H-section column, with gages close to the load axis, has reduced sensitivity to bending moments.

Although column spring elements have been, and continue to be, widely applied in high-capacity load cells, their popularity is evidently on the wane. In many cases, a properly designed shear-web spring element can offer greater output, better linearity, lower hysteresis, and faster response.

Note: Readers are reminded that many of the load cell configurations illustrated in this monograph (and countless others) have been the subject of one or more patents, some of which may currently be in force. It is therefore recommended that the patent literature be consulted prior to undertaking the design or manufacture of a load cell.

3

Transducer Spring Materials

Selection of the spring material deserves very careful attention because of its effects on both performance and cost.

Properties

We have previously identified the *spring element* as the most critical mechanical component of any strain gage transducer. The load-reacting element is referred to as a spring, irrespective of its configuration, because it should exhibit the qualities of a high-precision spring. These include: linearity in deflection-versus-force, low hysteresis (essentially identical behavior for both increasing and decreasing forces), low creep, and low stress relaxation. Other than the unusual configurations which often characterize transducer spring elements, their principal distinguishing feature from conventional springs is in the respective deflection magnitudes. The classical concept of a spring, such as a leaf or coil spring, normally involves rather large deflection at rated load. In contrast, the deflection of a typical transducer spring element is only a few thousandths of an inch (say, 0.1 mm or less), even when the physical size of the transducer is quite large. Thus, the transducer element is, in effect, an extremely low-compliance, high-precision spring.

Though a transducer spring element must behave like a precision spring in most respects, it must also have a physical form which provides areas where the strain is high enough and sufficiently uniform to serve as sites for strain gage placement. Since it is usually necessary that total transducer deflection be minimized, a great deal of attention is given the element design to ensure that stress levels (and deformations) are as low as possible throughout the structure, except at gage locations.

One basic reason for minimizing the compliance of the spring element is the usual requirement for the highest practicable natural frequency of the transducer itself. Another reason stems from the fact that the mass of whatever loads the transducer is necessarily in series with the spring. In many applications (e.g., weighing), it is preferable to maximize the natural frequency of the resulting mass/spring system.

Selection of the transducer spring material deserves very careful attention because of its effects on both performance and cost. The selection process involves three main groups of considerations; namely, the mechanical properties of the material, its thermal properties, and a number of characteristics which can be referred to collectively as "manufacturing considerations".[2] All three topics are discussed in the following sections of this chapter.

Mechanical Properties

One of the more important mechanical properties in determining transducer accuracy is the elastic linearity of the material. Ideally, for the range of stress corresponding to the rated load of the transducer, the material should have a perfectly linear relationship between stress and strain. In addition, for many transducer designs, it is preferable that the slope of the stress/strain relationship be the same in compression as it is in tension. While nonlinearities in transducer output can arise from a number of other sources in addition to the material behavior, greatest accuracy will ordinarily be achieved by starting with a highly linear spring material.

The material should also be selected for purity of elastic behavior over the working stress range. Hysteresis, for instance, should be low so that sensibly the same strain is produced for the same stress during both the increasing and decreasing phases of any load cycle. Similarly, when any fixed load is applied, and held constant, the strain should remain essentially constant with time. In other words, the material should exhibit minimal creep. It should also be correspondingly free of stress relaxation at constant deformation. A further selection criterion is the constancy of the foregoing properties with respect to cyclic load history.

---

[2]Chemical properties are not considered in this discussion of spring materials. Corrosion resistance, for example, may or may not be important in a given transducer design. Many transducers are hermetically sealed, and those that are not are usually given a high degree of protection against atmospheric corrosion.

Conventionally cited structural properties of metals, such as the proportional limit, yield strength, and ultimate strength, do not bear directly on the performance of the material as a transducer spring element. These terms are relevant only at strain levels considerably beyond those developed in transducers at their rated loads. The structural properties of the spring material may be very important, however, with respect to its behavior when overloaded. Load cells, for instance, are usually required to sustain loads much higher than the full-scale load before failure occurs. This type of requirement necessitates consideration of the yield strength and related properties. If catastrophic failure under the prescribed overload is to be avoided, the material must display enough ductility to prevent fracture in regions of stress concentration such as threaded connections. It is for the latter reason that steel spring elements are seldom found with hardness levels greater than about Rc50. The alloy in a harder condition may be a slightly better spring material, but it will also be more prone to brittle failure.

In a well-designed spring element, the fatigue strength is usually a secondary consideration, even when the load cell is intended for high-cycle applications. This is because the fatigue susceptibility of the strain gage is ordinarily the limiting constraint. With the greatest strain in the spring occurring at the gage location, and with this level intentionally limited to assure long cyclic gage life, there should be no danger of fatigue failure in the element. An exception to this general principle is found in "universal" load cells — those designed for both tension and compression forces. When intended for applications with large numbers of reversed load cycles, the spring elements require special care in mechanical design and material selection to avoid fatigue failure. Such "fatigue-rated" transducers are usually designed for a reduced output factor (mV/V at rated load) to allow lower working strain levels in the structure and at the gage locations.

Not mentioned up to this point has been the elastic modulus of the spring material. The modulus can be a very important factor in material selection under certain conditions, depending primarily on whether the load rating of the transducer is high or low. Since this topic is linked with a variety of design and manufacturing considerations, it is treated separately in a later section.

Thermal Properties

The preceding discussion of mechanical properties was implicitly limited to "typical" transducer working conditions, which might be described as room temperature ±75° F, or ±40° C. Over such a temperature range, the mechanical properties do not usually change enough to affect material selection. When a transducer is designed for use at a very high or low temperature, however, special consideration must naturally be given to the material properties at the operating temperature. With extreme temperatures, this will severely limit the range of materials suitable for the application.

Beyond the foregoing, there are other temperature-related characteristics of the spring material which can strongly influence transducer performance. The more important of these are treated here as "thermal properties".

The heat transfer characteristics of the spring element constitute a major factor in determining the accuracy and repeatability of a transducer. This is because temperature gradients within the element (particularly those which are unsymmetrical with respect to the gage locations) can cause erratic and irreproducible perturbations in the transducer output. As a result, the spring element should always be designed with its heat flow characteristics in mind, as well as its mechanical features. For any design, under any combination of internal and external heat inputs, temperature gradients within the element will vary inversely with the thermal conductivity of the material. Thus, thermal conductivity is a significant property which deserves attention when selecting a spring material.

The expansion coefficient of the material, although one of the more widely published thermal properties, need not be a major consideration in spring material selection. The first-order effects of thermal expansion on transducer output are normally eliminated by using self-temperature-compensated strain gages in full-bridge circuits. There are also higher-order effects, caused by the dimensional changes in cross sections, moment arms, and similar geometric parameters. These are typically in the order of 0.1 to 0.2% for a temperature excusion of 100° F, or 60° C. Their effects on transducer output, however, are indistinguishably commingled with other thermally induced deviations such as those described in the following paragraph. Any corrective measure must be applied to the net output deviation with temperature due to all effects combined.

Most materials experience a reduction in elastic modulus with temperature. For metals, there is typically a 1 to 3% decrease in modulus per 100° F, or 60° C, temperature rise. This effect, by itself, would cause the "span" of a transducer (full-scale output signal) to increase as the temperature rose. While a special "constant-modulus" alloy (such as found in some of the more accurate mechanical scales using springs) might be considered for transducer use, it would not completely solve the problem. The reason is that the gage factor of a strain gage also changes with temperature; and this, too, directly affects the span of the transducer. Fortunately, there now exist several practical and highly effective methods of compensation for controlling thermally caused span changes. With these methods, overall transducer calibration can be held within very close limits over relatively wide temperature ranges.

Manufacturing Considerations

In addition to the mechanical and thermal properties of a spring material, there are other characteristics, broadly related to manufacturing, which must be considered in making a selection. One of the most important of these is machinability. Modern transducers frequently employ spring elements with extremely complex configurations. Since the spring is preferably fabricated in one piece from a solid billet, intricate precision machining is a common requirement. In such cases, the machinability of the alloy can significantly affect both the performance and the cost of the finished transducer.

Hand in hand with machinability is the desirable alloy quality of freedom from distortion during the hardening process. Many potentially suitable spring alloys are machinable only in the soft condition. If high-temperature heat treatment is subsequently required to develop the desired mechanical properties, serious distortion of the original shape can occur. This is particularly true of alloys which require a rapid oil or water quench from a high heat-treatment temperature. Consequently, such alloys are seldom used when the element configuration involves a complex pattern of both thick and thin sections.

In some transducer designs it is necessary to attach flexures or other appendages to the spring element, or to fabricate the element itself from more than one piece, by welding or brazing. In such cases, the latter properties of the material will, of course, affect its suitability for the application. Another factor which often bears importantly on the choice of spring material is its commercial availability in a usable form and shape, and in quantities consistent with projected production needs. Material cost, although usually a secondary consideration, may also influence the decision.

Because no single spring material is equally suitable for all types of applications, the selection process necessarily involves compromise — among mechanical properties, thermal properties, and manufacturing or production considerations. A knowledgeable transducer designer arrives at the most appropriate spring material for a particular application by weighing the relative priorities to be given the foregoing properties against a variety of practical criteria which will determine the viability of the transducer. Examples of the latter are: load capacity, accuracy requirements, physical size or shape limitations, operating environment, expected production volume, and cost constraints.

High- Versus Low-Modulus Materials

The required characteristics peculiar to a transducer spring element give special prominence to the modulus of elasticity in the material selection process. It is convenient, for this purpose, to classify the various materials suitable for transducer elements as either high-modulus (the steels) or low-modulus (aluminum and other non-ferrous alloys). While there is no sharp dividing line in the applications, high-modulus materials are generally used for high-force transducers, and low-modulus materials for low-force units. As described in the following paragraphs, there are several reasons for selecting a material on this basis.

It was noted earlier that a transducer spring element should be characterized by very low overall compliance. At the same time, however, it must contain local regions where the strain is high enough (typically 1000 to 1500$\mu\epsilon$) to serve as strain gage installation sites. In the case of high-capacity load cells, there is usually no particular problem in meeting these requirements with a well-proportioned steel element. As rated loads decrease, however, the material thickness in sections which receive strain gages must be progressively reduced to develop the required strain level. Very thin sections at the gage sites can introduce a variety of effects which tend to degrade transducer performance.

Thinner material sections give rise to increasingly serious reinforcement of the spring element by the strain gage itself. This can be particularly noticeable in bending elements. As the beam thickness decreases, an ever greater fraction of the overall section at the gage site consists of strain gage; and the grid of the gage is relatively farther from the beam surface. The material between the grid and the beam surface is a plastic matrix, composed of the gage backing and the bonding adhesive. The result is that the transducer element becomes substantially reinforced by plastic, and the composite structure begins to display excessive creep, seen as a time-dependent relationship between gage output and applied force. Thin sections at the strain gage locations are also detrimental to heat transfer within the element. Most of the heat generated in the gage must flow from the grid to the element surface; and from that surface, through the section, to adjacent sections of greater mass. With unduly thin sections at the gage sites, this impedance to the heat flow can cause significant temperature gradients and a correspondingly higher temperature in the gage. Such a condition not only has a direct effect on gage output, but it can also alter the local strain field by developing thermal stresses in the section, creating a particularly troublesome performance problem.

Still other problems with thin sections can occur in fabricating the spring element. Many complex element designs require intricate machining procedures; and it is not easy to machine a very thin web without producing local deformations or residual stresses in the section. It is extremely difficult to accurately measure machining deformations, and still more difficult to do anything about them when they are found, except to scrap the part. Residual stresses which are not accompanied by measurable distortion of the section are almost impossible to detect by nondestructive methods. When present, however, residual stresses can seriously affect transducer performance. They also tend to cause considerable unit-to-unit variation in performance among otherwise identical elements.

One of the most effective means for achieving greater section thicknesses in low-capacity spring elements is the use of a material with a low modulus of elasticity; for example, an aluminum alloy. Although aluminum alloys are not characteristically thought of as spring materials, they function very well in this role when the element is properly designed. As a result, aluminum-alloy spring elements are widely used in low-capacity transducers.

Material Selection

Described in the preceding section were the principal attributes which need to be considered in the process of selecting a transducer spring material — namely, mechanical properties, thermal properties, and a group of additional factors classified as "manufacturing considerations".

Linear-elastic load response, with minimal hysteresis, was identified as one of the most desirable mechanical properties for a spring material. Among thermal properties, the thermal conductivity was highlighted as a key factor affecting transducer accuracy and stability. Because practicality in fabrication is always important, machinability and distortion-free hardenability ranked high among the manufacturing considerations.

The elastic modulus of the material was selected for special attention because of its important role in determining spring element section dimensions with respect to load rating. While steel alloys make excellent materials for high-capacity spring elements, they are often unsatisfactory for low-capacity units. In the latter case, to achieve suitable strain levels in reasonably thick and easy-to-machine sections, it is generally necessary to use low-modulus materials such as the aluminum alloys.

This section presents a survey of "candidate" materials for transducer spring elements; i.e., materials which are commonly used, as well as some which might be considered, for this purpose. The candidates range from materials which are well suited to the function, and can be highly recommended, to those which should never be used except as dictated by special application requirements. The latter are included because their deficiencies help clarify the criteria for selecting superior materials.

Material Selection Guide

The table on page 37 is intended as an aid in selecting a transducer spring material. It will be noted that materials in the table are divided into three categories: high-modulus, low-modulus, and "other", or miscellaneous. Given first for each material are the principal mechanical and thermal properties relevant to spring element design. These are followed by a series of comparative ratings of the material for other, less readily quantified, characteristics such as linearity, hysteresis, etc. The ratings, or "figures of merit" are on a scale from 1 to 10, with higher numbers indicative of superior qualities.

It should be noted that the tabulated properties data represent *nominal*, room-temperature values for the material in the "as-normally-used" condition. These data were drawn from, and cross-checked with, a large number of sources — which sometimes show serious disagreement on various material properties for what is presumably the same material. The data are believed to be quite accurate, however, and suitable for the purpose of material selection. Tabular quantities are expressed in U.S. customary units, but metric/S.I. conversion factors are provided in the footnotes. The qualitative ratings, which are intended simply for between-material comparisons, reflect practical experience in the design and manufacture of transducers. Since such experience obviously does not encompass every material in every conceivable application, the "figures of merit" are intended as only rough guides.

High-Modulus Alloys (Steels)

Types 4140 and 4340 alloy steels have been used for many years in spring elements of high-capacity load cells. These steels represent a particularly good choice when the elements are large, in which case the necessary heat treatment for hardening is not likely to cause serious distortion. The precipitation-hardening stainless steels, and the maraging steels, are currently the most popular high-modulus spring materials. They are easily machined in the unhardened state, and can be brought to their final properties with a low-temperature heat treatment that results in minimal distortion. The corrosion resistance of the precipitation-hardening stainless steels is also a great advantage in many transducer designs. Probably the most serious limitation of these alloys is their low thermal conductivity, coupled with a tendency for significant differences in physical properties among different melts. Type 17-4 PH, for instance, although widely used, is one of the most variable alloys from melt to melt. Type 15-5 PH, on the other hand, has properties almost identical to 17-4 PH, but reputedly has better lot-to-lot consistency, as well as improved mechanical properties in the transverse direction.

In some cases, the final choice of the spring-element alloy may be largely dictated by the availabilities of the various alloys in the required size and form. Type 17-7 PH, for example, is an excellent material for this purpose, but is supplied primarily in plate and sheet form.

Low-Modulus Alloys

Aluminum alloys are the most widely used low-modulus materials at the present time. Of several potentially suitable alloys, types 2024-T4 and 2024-T351

Materials Selection Guide for Transducer Spring Elements

MATERIAL	PROPERTIES (in "as-normally-used" condition)*								RELATIVE *FIGURE OF MERIT*†										REMARKS
	Modulus of Elasticity, psi x 10^{-6} [1]	Yield Strength (0.2% offset) psi x 10^{-3} [2]	Thermal Expansion Coeff., ppm/°F [3]	Density, lb/in³ [4]	Max. Recommended Temperature, °F [5]	Thermal Conductivity $\frac{\text{Btu}}{\text{hr} \cdot \text{ft}^2 \cdot (°\text{F/in})}$ [6]	Elongation, % [7]	Hardness (R_c, unless noted)	Thermal Conductivity	Linearity	Hysteresis	Creep/Relaxation	Material Cost	Machinability	Hardening Distortion	Corrosion Resistance	Welding or Brazing Suitability	Lot-to-lot Consistency	
High-Modulus Alloys (AISI Steels)																			
4140	30	200	6.5	0.283	+400	270	12	42-45	3	8	8	7	9	5	3	1	5	6	very good steel alloy
4340	30	210	6.3	0.283	+500	260	12	43-46	3	8	8	8	8	5	3	1	5	8	excellent steel alloy
18 Ni (250), maraging	27	245	5.6	0.289	+600	140	7-10	48-49	2	8	8	7	4	5	8	2	8	5	good, but seldom used
410 stainless	29	140	6.0	0.280	+400	170	15	39	2	7	7	6	7	4	2	6	2	6	seldom used
630 stainless (17- 4 PH)	28.5	185	6.0	0.280	+400	135	14	41-44	2	7	7	7	6	6	8	7	7	3-4	widely used
631 stainless (17-7 PH)	29	220	5.7	0.276	+400	130	6-7	46-48	2	8	8	7	6	6	8	7	7	5	widely used
632 stainless (PH 15-7 Mo)	29	220	5.0	0.277	+500	110	6	47	2	8	8	8	4	6	8	7	7	5	excellent, but seldom used
S15500 stainless (15-5 PH)	28.5	185	6.0	0.280	+400	135	14	41-44	2	7	7	7	6	6	8	7	7	5-6	improved version of 630
Low-Modulus Alloys																			
2014-T6 alum. alloy	10.6	60	12.4	0.101	+200	1075	10	B 135	8	6	7	6	5	8	9	3	3	6	good
2024-T4/T351 alum. alloy	10.6	46	12.6	0.101	+200	840	18	B 120	7	6	7	6	5	8	9	3	3	6	good, widely used
2024-T81 alum. alloy	10.6	65	12.6	0.101	+250	1040	8	B 128	8	7	8	7	5	8	9	3	3	6	best of aluminum alloys
6061-T6 alum. alloy	10.0	40	12.9	0.098	+150	1160	14	B 95	8	5	4	4	6	7	8	4	5	5	fair performance
7075-T6 alum. alloy	10.4	70	12.9	0.101	+100	840	10	B 150	7	7	6	6	5	8	8	3	2	5	poor at elevated temperatures
BeCu 25 (C17200)	17	170	9.3	0.298	+250	750	2-3	40-42	7	8	8	8	2	5	9	3	4	7	excellent, but costly
6 Al 4V titanium	16.5	165	4.9	0.160	+300	50	12	40	1	7	7	7	2	3	2	8	4	5	used in special applications only
Miscellaneous Materials																			
AISI 304 stainless steel	28	150	9.6	0.290	+250	110	4-5	39	2	5	4	4	7	3	(B)	8	5	6	poor spring material
Ni-Span-C alloy	28	180	4.3	0.294	+250	90	9	37-38	1	8	8	8	3	3	4	3	5	7	very good, but costly
AZ31B magnesium alloy	6.5	32	14.5	0.064	+100	650	15	B 73	6	3	2	2	6	8	(B)	1	7	5	very poor spring material
Ceramic, Al_2O_3 (99+%)	55	60-70	3.3	0.141	+500	260	(A)	Mohs:9	3	7	8	8	1-3	1	(B)	10	—	7	special applications only
Epoxy-glass laminate (high-pressure, low resin content)	5-7	70-100	5-8	0.06-0.07	100-200	2	—	—	0	5	4	3	5	3	(B)	4	—	3	special, low-accuracy applications only

**The mechanical and physical properties given in this table have been carefully selected from a wide range of sources as representative for typical transducer applications and environments.*

†*Numbers tabulated for* FIGURE OF MERIT *represent relative ratings on a scale of 0 to 10, with higher numbers indicative of more desirable qualities. The ratings are based on experience, judgment, and current practices in the transducer field.*

[1] for GPa, multiply by 6.9
[2] for MPa, multiply by 6.9
[3] from 75°F to 212°F; for ppm/°C, multiply by 1.8
[4] for g/cm³, multiply by 27.7
[5] transducer applications; for °C, subtract 32 and multiply by 5/9
[6] for watt/m·°C, multiply by 0.144
[7] in 2-in *(51-mm)* gauge length

(A) no ductile range
(B) not hardenable by heat treatment

are the most readily available and commonly specified. Perhaps the best all-around aluminum alloy for transducers is 2024-T81, which is basically 2024-T4 with an aging treatment in the neighborhood of 375° F *(190° C)* to put it in the T81 condition.

From the standpoint of overall performance, beryllium copper is probably the best of the low-modulus materials. It is expensive, however, and special precautions must be taken during machining because of the health hazards associated with beryllium. Several titanium alloys make good spring materials; but they are expensive, as well as difficult to machine and to harden without distortion. Their greatest performance drawback is their extremely low thermal conductivity.

Other Materials

Magnesium alloys offer several interesting properties for low-modulus applications. Unfortunately, the currently available magnesium alloys show very poor spring characteristics, along with serious time-dependent effects; and, for these reasons, are rarely used.

The 18-8 stainless steels are readily available, and are characterized by excellent corrosion resistance. They are not hardenable by heat treatment, however, and thus find only very limited application.

Ni-Span-C is a nickel alloy developed specifically for use in precision springs. Its chief virtue in these applications is that, with proper heat treatment, its elastic modulus can be made essentially invariant with temperature over a fairly wide range. But, as noted in the previous section, this characteristic is of much less importance in strain gage transducer elements. Although Ni-Span-C is an outstanding spring material, its transducer applications are limited by its high cost and poor machinability.

Fiber-reinforced plastics are increasingly used in structural applications today, although their plastic content creates time-dependent effects which would normally preclude their use as transducer elements. They are occasionally specified, however, when the transducer is not a force-sensitive type. A cantilever beam, for example, can be adapted to measure the displacement of a plunger by gaging the beam to sense bending. In this case, changes in stiffness due to creep and relaxation would be much less important; and high-glass-content laminates are occasionally found in such applications. As properties of fiber-reinforced composites continue to improve, their application in low-force transducers of low or moderate accuracy can be expected to increase correspondingly.

4

The Strain Gage System

Since transducer behavior can be observed to a resolution better than one part in 20 000, the strain gage system must be selected and installed with the utmost care.

Gage Selection

Previous sections have dealt principally with design considerations and material selection for load-cell spring elements. It is now appropriate to focus on the selection and installation of the strain gage system for transducers. Much of the background information for this discussion can be found in Measurements Group Tech Note TN-505, *Strain Gage Selection.* Although intended primarily for stress analysis applications, TN-505 is recommended as a very useful reference for the transducer-oriented material presented here.

In contrast to the situation for stress analysis applications, the strain gages installed on a transducer can readily be calibrated against known physical standards — dead weights, for example, or a previously calibrated transducer. The existence of precise standards and sensitive electronic instrumentation allows the constructor of a transducer to quantify its performance to a very high degree. It is possible, in fact, to observe transducer behavior to a resolution better than one part in 20 000. This corresponds, in effect, to detecting a strain of $0.05\mu\epsilon$ or less on the surface of the spring element. Clearly, the resolution of such small dimensional changes requires that the strain gage system be selected and installed with the utmost care.

The gage selection procedure for transducer applications is similar to that outlined in Tech Note TN-505. For most cases, the preferred sequence is:

1. Gage length
2. Pattern geometry
3. Gage series
4. Grid resistance
5. S-T-C number
6. Optional features

Selection criteria for the above parameters are given in the following sections.

Gage Length

Practical transducer design considerations dictate that the strain gage be mounted in the area of highest strain. This produces the maximum allowable signal output consistent with the strength of the spring element material. At the same time, it minimizes the likelihood of fatigue failures originating at points of stress concentration outside of the carefully designed gaging site. Maximizing the strain at only the gage location also minimizes the spring-element deflection and its detrimental effects on natural frequency and linearity.

Consider, for instance, the simple binocular-style load beam in Fig. 4.1b. The two holes are designed to provide a flexural sensitivity significantly higher (with respect to beam deflection) than is possible in a straight, uniform beam. The holes produce localized strain concentrations on the gaging surface as sketched in Fig. 4.1a. Dimension X represents the area of maximum, nominally uniform strain. Based on the foregoing premise, the active gage length (including endloops) should not be greater than X.

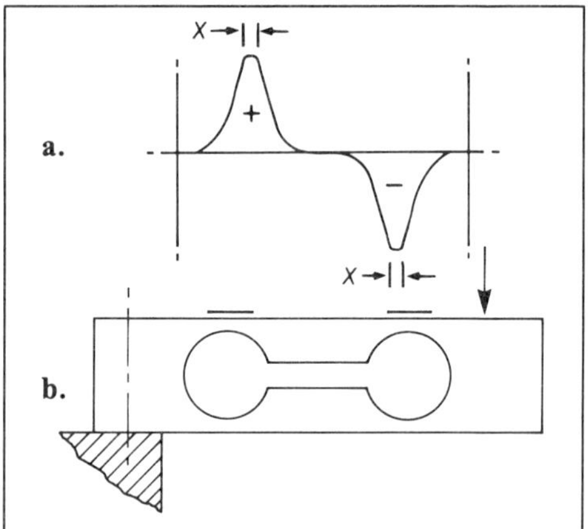

Fig. 4.1 — Approximate strain distribution and gage locations for binocular spring element.

For this and most other transducer designs, a good general rule of thumb is that the strain under the gage grid should not vary by more than 10 to 15% from the absolute maximum. This condition will give the highest possible output while maintaining essentially uniform strain under the grid. Uniformity of grid strain maximizes the fatigue life of the gage, and reduces variations in transducer creep under load. In the case of transducer designs which have inherently large areas of uniform strain — e.g., constant-stress beams and shear-web designs — gage length selection becomes primarily a matter of available space, power dissipation requirements, and gage cost. These factors are usually optimized by selecting gage lengths in the range from 0.060 to 0.125 in *(1.5 to 3.2 mm).*

Grid Geometry

Once the gage length has been determined, catalog selection of the gage pattern is ordinarily the next step. For a gage length of 0.060 in *(1.5 mm)*, there are typically several pattern selection criteria to be considered:

Grid Width — In general, wide patterns are preferred over narrow grid designs because the larger grid area provides better power handling capability. This benefit must always be weighed against cost, however, since narrow patterns usually have a price advantage. Disregarding power-dissipation considerations, most transducer designs perform equally well with narrow or wide patterns (Fig. 4.2) — although there are some significant exceptions to this generality which must be reserved for future treatment.

Multiple-grid patterns — For many types of transducers, gaging locations are such that two gages are installed in close proximity to each other in one or more locations. Fig. 4.3 is representative of several configurations in which two gages, connected as adjacent bridge arms, are required at both locations *A* and *B*. In this and similar applications, there are several advantages in having both grids on a common backing as shown in Figs. 4.4a and 4.4b:

a. Gage installation and alignment are faster and less subject to error.
b. Since each gage pair operates as a half-bridge, with one gage in tension and the other in compression, the grids can share a common solder tab, thus reducing the wiring time.
c. The two grids are produced from the same local area of foil, and therefore have nearly identical thermal properties. The result is inherently a closer tracking of bridge balance with temperature change. It should be noted that this important feature relies on having the two grids (of a common backing) in *adjacent* bridge arms.

Another application of multiple-grid patterns occurs in simple bending-beam transducers such as that shown in Figure 4.5a. Generally, the same advantages of the multiple-grid pattern accrue here, except that the improvement in zero-shift with temperature is not realized because the grids in each pair are necessarily connected in *opposite* bridge arms.

Tab geometry — Some gage patterns are offered in more than one solder-tab configuration; for example, with both tabs at one end, or with a tab at either end

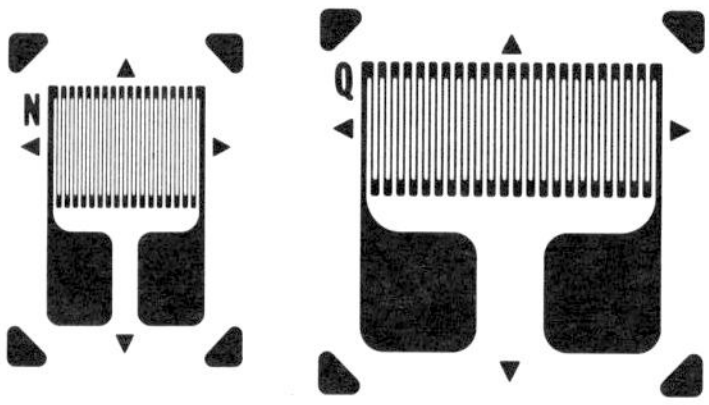

Fig. 4.2 — Typical single-element grid patterns for transducer applications.

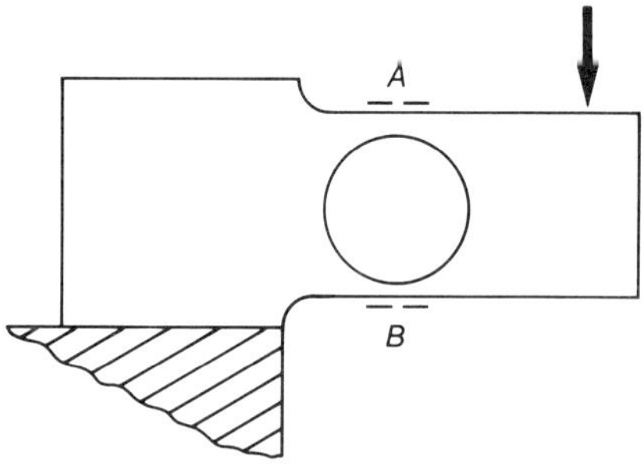

Fig. 4.3 — One of the many spring element designs which can advantageously enploy paired strain gages on a common backing.

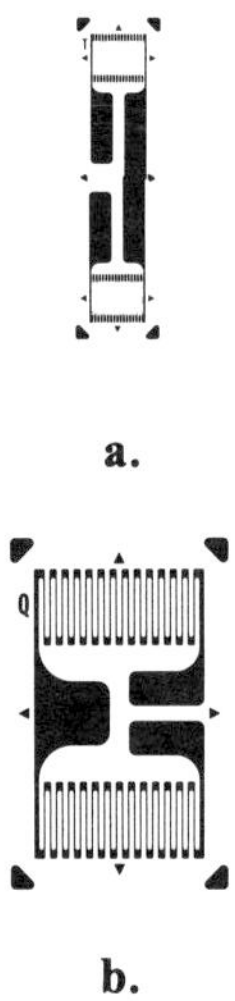

Fig. 4.4 — Dual-grid strain gage patterns designed for use on spring elements with tensions and compression gage areas in close proximity.

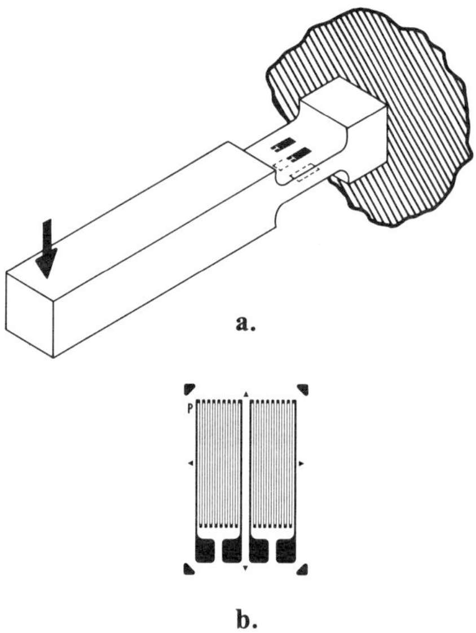

Fig. 4.5 — Side-by-side dual grid gage patterns for bending beam applications. A corresponding pair is installed on the under side of the beam.

(Fig. 4.6). The choice between the two tab arrangements is often made on the basis of wiring convenience, but other factors also require consideration. If a significant temperature gradient can exist along the length of the gage, the opposed tab configuration in Fig. 4.6b can create serious problems with thermal EMF generation at the leadwire/solder-tab junctions. This becomes most important when d.c. instrumentation is used, and when the strain gage alloy (constantan, for instance) has a high thermal EMF against copper.

Gage Series

With the gage length and pattern established, the next step in the selection procedure is normally to choose the gage series best suited for the intended application. The term *gage series* refers generally to the combination of grid alloy and backing (carrier) embodied in the gage construction. For simplicity in presentation, the backing and alloy are treated separately here; but the final selection must be made directly from the gage catalog, since not all backing/alloy combinations may be offered.

Gage backings — The backing of a strain gage performs a variety of functions:

a. It provides a carrier for the delicate sensing foil so that the gage can be handled and installed without damage.

b. It insulates the conductive metal from the substrate material.

c. Most importantly, it acts as a strain-transfer medium through which the strain in the spring element is transmitted to the sensing foil. The capacity of the backing to perform the transfer function accurately and repeatably over the prescribed range of temperature and strain conditions is the primary selection criterion for a transducer gage backing. The capacity for strain transfer depends on the rigidity or stiffness of the backing as manifested by its shear modulus. Since the shear modulus of a plastic tends to fall with rising temperature, the higher-modulus backings generally perform better at elevated temperatures. High-modulus backings, however, are usually less flexible and more brittle than their general-purpose counterparts. At the other extreme, some plastics which would otherwise be suitable for gage backings perform poorly because of excessive viscoelastic behavior.

Three broad types of strain gage backings are commonly used in transducers. These are: polyimides, epoxies, and reinforced epoxies. Polyimides represent the most rugged and flexible (virtually unbreakable) backing materials currently in use. Historically, polyimide backings have been relegated to lower accuracy transducer use because they exhibited greater viscoelastic (creep) behavior than the epoxies. Recently, a new family of polyimide gages has been introduced by the Micro-Measurements Division that retains the favorable rugged characteristics of the earlier polyimides while providing substantial improvements to viscoelastic properties. The resulting reduction in gage backing creep has brought about widespread acceptance of the new materials in the highest precision transducer applications.

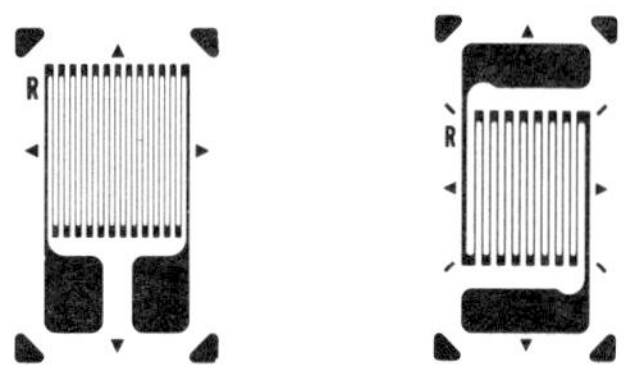

a. b.

Fig. 4.6 — Representative solder tab variations intended for different installation and operating requirements

Currently, the most widely used backings for precision transducers are made from specially formulated unfilled epoxy. They are relatively high in shear modulus and low in "cold-flow" tendency; and they can be produced in thin backing films [typically, 0.0009 in *(0.023 mm)*]. This combination of properties yields a dimensionally stable backing characterized by extremely low creep. With proper grid selection, room-temperature creep can be held to 0.02%, or better, of full-scale output. In common with all unfilled plastic resins, however, the creep increases with temperature; and precision-transducer applications of these backings are generally limited to temperatures below 200° F *(95° C)*.

Both high-temperature stability and creep performance can be significantly improved in epoxy gage backings by the inclusion of glass fibers in the backing. A random-fiber matrix improves gage strength and flexibility, and permits the use of rigid binder resins that would be too brittle for use in the unfilled form. The effective elastic modulus of the gage is also increased significantly by the inclusion of the glass fibers. This can cause transducer creep problems due to structural reinforcement of the spring element by the gage. The effect becomes noticeable with thin spring element sections in bending; and particularly so when the element is made from a low-modulus material such as an aluminum alloy. Since a significant fraction of the spring-element bending moment is carried by the gage, the effect is manifested as transducer creep.

Grid alloys — The second component in gage-series selection is designation of the strain-sensing grid alloy. Four alloy types are commonly used in modern transducers: constantan, Karma, Isoelastic, and platinum-tungsten. Since not all backing/alloy combinations are mutually compatible, the strain gage catalog must be consulted for series selection.

Constantan, a copper-nickel alloy, is the most popular alloy for transducer gages. It possesses exceptional linearity over a wide strain range, and is readily manufactured in an extensive array of grid configurations. It is also easily solderable, and is subject to precise temperature compensation when used in conventional Wheatstone bridge circuits. The primary limitation in precision transducer applications is a slow, irreversible drift in grid resistance when exposed to temperatures above about 150° F *(75° C)*. The effect of the drift is considerably reduced in full-bridge circuits, however, since all four gages drift in the same direction at similar rates. Because the drift rate increases exponentially with temperature, constantan is not recommended for transducers operating continuously at temperatures above 200° F *(95° C)*. Drift at lower temperatures is not otherwise damaging to transducer operation; and constantan can be used in a variety of such applications where it is practicable to re-zero the bridge to compensate for the slight drift in grid resistance. Constantan also displays a very high thermal EMF in contact with copper, and this may require special consideration in certain cases.

Karma is a nickel-chromium alloy, used in a variety of modified forms for strain sensing. Like constantan, it displays extremely good linearity over a wide strain range. It has greater resistivity than constantan, making higher grid resistances feasible. This alloy can also be temperature-compensated but not quite to the high degree possible with constantan. A major advantage of "K-alloy" is its improved resistive stability, particularly at elevated temperatures. The fatigue life of the alloy is also much greater than for constantan, providing an important benefit in dynamic transducer applications such as "fatigue-rated" designs.

Certain specially modified Karma alloys have a unique capability for compensating transducer span changes with temperature. This characteristic is achieved by controlled alloying and processing to produce an adjustable negative change in gage factor with temperature rise. With testing, the decrease in gage factor can be closely matched to the reduction in transducer spring modulus with temperature. The result is a transducer with a span which is essentially independent of temperature. This method of "modulus compensation" is normally limited to use by commercial transducer manufacturers because of the extensive testing and alloy lot qualification required.

Karma alloy gages are not without certain drawbacks. They are more expensive to manufacture, for instance, and the nickel-chromium alloy is difficult to solder without special soldering aids. For this latter reason, all open-faced K-alloy *Transducer-Class®* gages are supplied with copper soldering pads as shown in Fig. 4.7.

Isoelastic alloy offers exceptionally good fatigue life and a gage factor about 50% higher than constantan and the Karma alloys. It has only limited use in transducers, however, because it cannot be temperature-compensated. Isoelastic can sometimes be used in full-bridge circuits for low-accuracy static measurements; but it is normally restricted to dynamic transducer applications because of its poor zero stability with temperature variation. It is also difficult to solder, and requires special procedures similar to those for Karma alloys.

Platinum-tungsten alloys, like Isoelastic, find their primary use in dynamic measurements. These alloys are not easily temperature-compensated, even in full-bridge circuits. With a gage factor approximately two times greater than constantan and Karma, and with very good fatigue life, platinum-tungsten gages are used almost exclusively in "fatigue-rated" dynamic transducers. In such applications, the spring-element strain can be kept low enough to ensure long transducer life without undue sacrifice in output signal. Since platinum-tungsten is a difficult material from which to fabricate strain gage grids, the available gage selection is very limited, and the prices are high.

Grid Resistance

The strain gage resistance is an important element in transducer design, primarily because of its combined effects on output signal amplitude and power dissipation. Transducers are commonly classified in sensitivity according to their full-scale output in mV/V (millivolts of output signal per volt of circuit excitation). In a typical 2mV/V transducer, a 10V excitation source thus produces a 20mV output signal under full rated load. Obviously, the signal-to-noise ratio of such a unit can be noticeably improved by increasing the excitation voltage to 15, or even 20, volts.

The improvement in signal quality is not "free", however, since the power dissipated in a strain gage rises as the square of the excitation voltage. This may cause problems in battery-operated transducers, for instance, where energy consumption is usually a critical consideration. Of broader significance, it may also cause problems in transducer stability and accuracy due to thermal effects. All of the power consumed by the gages in a transducer must be dissipated as heat, flowing through the spring element to the structure of the unit. Depending on the material of the spring element, and its geometric configuration, along with the size and disposition of the gages, the maximum allowable power dissipation is ordinarily limited by the accuracy requirements.

In practice, many of the mechanical design parameters for a transducer are often largely fixed by the nature of the application. An attractive alternative for increasing the output signal without adding to the power dissipation is to use a higher resistance gage in conjunction with increased excitation voltage. With this consideration in mind, it is usually advantageous to select the highest practical gage resistance. Since most gage types are available in both 120- and 350-ohm versions, the latter is then preferred for transducers. In some instances, 1000-ohm gages are employed for the same reasons. Micro-Measurements also offers strain gages with resistances up to 5000 ohms in a grid size of only 0.050 x 0.060 in *(~1.3 x 1.5 mm)*, as shown in Fig. 4.7.

Fig. 4.7 — 5000-ohm transducer strain gage with copper soldering pads on gage tabs.

Detailed recommendations for excitation voltage — in terms of gage size and resistance, heat-transfer characteristics of the substrate, and accuracy expectations — are given in Measurements Group Tech Note TN-502, *Optimizing Strain Gage Excitation Levels.* As a rough guide, applicable to transducer spring elements with generous, symmetrically disposed paths for heat flow, the diagram in Fig. 4.8 illustrates the manner in which the recommended excitation voltage varies with grid area for different gage resistances.

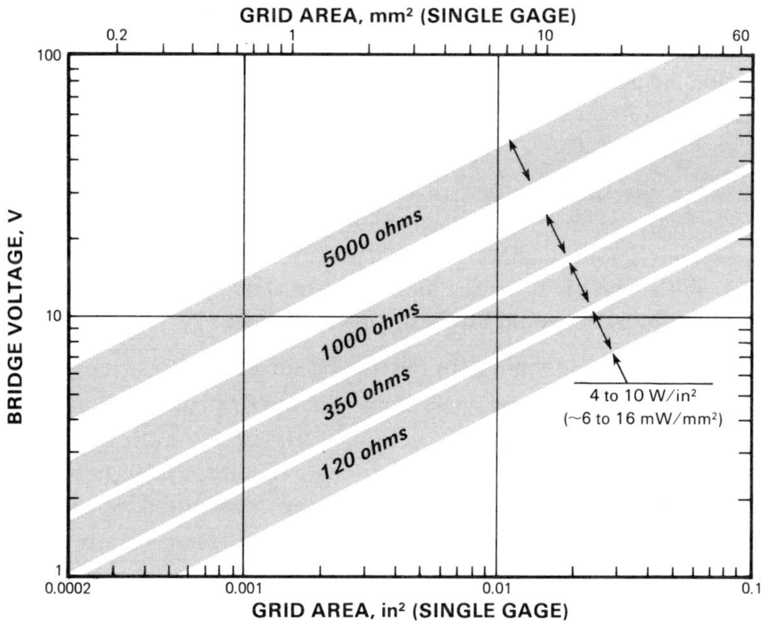

Fig. 4.8 — Bridge voltage as a function of grid area and gage resistance for power-density range of 4 to 10 W/in² *(~6 to 16 mW/mm²).*

Self-Temperature Compensation

The half-bridge and full-bridge circuits commonly used in transducers tend to provide inherent compensation for thermal output (apparent strain) if all gages have identical characteristics and sense identical temperatures. Since these conditions can never be precisely satisfied, best transducer accuracy will be achieved by selecting the self-temperature-compensation (S-T-C) number of the gages to match the (Fahrenheit) thermal expansion coefficient of the spring material.

An exception to this rule is sometimes made when the transducer is designed for operation in a special, limited, temperature range, as in cryogenic applications. Another exception occurs in the use of "modulus-compensating" K-alloy gages (described in the preceding section) which have specially tailored properties to compensate for variations in the elastic modulus of the spring element with temperature. When conditions such as these are encountered, the Micro-Measurements Transducer Applications Department should be consulted for recommendations.

Adhesive Selection

No single step in the manufacture of a transducer has more influence on its performance and longevity than strain gage bonding. Critical to this process is the selection of suitable adhesive. Improper selection or application of adhesive will surely degrade transducer performance, regardless of the quality of strain gage or spring material.

Criteria for selecting an adhesive are similar to those for the gage backing, since the adhesive is, in effect, an extension of the gage backing to the spring-element surface. The ideal adhesive would make the strain gage act as an integral and inseparable part of the spring element — without adding influences of its own. Such an adhesive would be characterized as follows:

- is easily prepared and applied to the mating surfaces.
- is capable of a strong, permanent bond, without difficult handling or curing procedures.
- provides a thin, void-free "glueline" with high rigidity in shear; and faithfully transmits the underlying strain field to the gage with negligible creep.
- maintains its properties throughout the specified temperature range and operating life of the transducer.

In practice, compromise is usually necessary when selecting an adhesive for a particular transducer application. Since certain characteristics are more important than others in different cases, several adhesive systems have been developed to serve the spectrum of needs. Among the adhesives offered by Micro-Measurements, those finding use in transducer construction (in varying degrees) are described below.

Cyanoacrylate Adhesive (M-Bond 200)

This is a single-component adhesive which transforms from a liquid to a solid when pressed into a thin film between the gage and mounting surface. It is very popular for routine stress analysis applications because of its fast cure and minimal clamping requirements (firm thumb pressure for 1 to 2 minutes is normally sufficient). M-Bond 200 is sometimes used in simple do-it-yourself transducers where high accuracy is not required. The adhesive is sensitive, however, to moisture and to certain solvents, and is also subject to long-term embrittlement. Thus, it is not recommended for use in a transducer which must serve for more than a few months. Creep effects are also less predictable than for the adhesive types that follow.

Epoxy Adhesives (M-Bond AE-10/15, 610, 600, 43-B)

Epoxies form the largest class of adhesives for strain gage bonding because of the range of characteristics available through proper formulation. Variations include filled adhesives, and two groups of unfilled adhesives classified as "100%-solids" and "solvent-thinned". Some epoxies are single-component, and others are two-component adhesive systems.[3]

Epoxy adhesives containing fillers, such as powdered aluminum oxide or silica, are often used in structural applications. The presence of fillers helps control the thermal expansion/contraction of the adhesive, and thus contributes to maintaining bond integrity over a wide temperature range. In strain gage installations, fillers also tend to increase the glueline thickness to 0.002 in *(0.05 mm)* or more, generally rendering the filled epoxies unsuitable for transducer applications. Except under rather rare circumstances, transducer gages should be installed with very thin gluelines — on the order of 0.0001 to 0.0005 in *(0.003 to 0.013 mm)*. This is done to improve strain transmission, reduce creep, and minimize reinforcement of thin spring-element members in bending.

Unfilled epoxy adhesives are capable of developing extremely thin, void-free gluelines, and thus represent the best choice for transducer installations. As noted above, two distinct classes of unfilled epoxies are formulated for strain

[3]The terms *single-component* and *two-component*, as applied to adhesives, are sometimes misunderstood. These descriptions refer only to the separate packaging of the curing agent(s) and epoxy resin. Either of these two compounds may actually consist of several constituents. Single-component epoxies are formulated with latent curing agents that require elevated temperatures to activate the curing mechanism. These adhesives have inherently long pot lives at room temperature, and thus do not require physical separation of curing agent and epoxy resin prior to use.

gage bonding: 100%-solids, and solvent-thinned adhesives. The term "100%-solids" means that the adhesive contains no ingredients in the form of volatile solvents which must be driven off before the adhesive cures. Micro-Measurements M-Bond AE-10 and AE-15 adhesive formulations are in this class. They both employ the same epoxy resin, but have different curing agents, Type 10 and Type 15.

M-Bond AE-10 and AE-15 differ most notably in the working time (pot life) available after the curing agent is mixed with the resin. AE-10 is the more reactive material, with a pot life of about 20 minutes, in contrast to 1-1/2 hours for AE-15. Because of its greater reactivity, AE-10 will cure sufficiently in six hours at room temperature for use in low-performance transducers. The preferred curing time is at least 24 hours at room temperature. AE-15 requires a higher temperature for complete polymerization; and curing for one hour at 175°F *(80° C)* is recommended as a minimum. This adhesive, however, develops higher ultimate properties than does AE-10. As a result, the AE-10 formulation is seldom selected for transducer gage bonding except where the elevated-temperature cure of AE-15 is impracticable.

Typical bondline thicknesses with AE-15 will normally be close to 0.001 in *(0.025 mm)*. This is somewhat thicker than ideal for high-precision transducers, principally because of creep effects. However, when used in conjunction with the general-purpose polyimide gage backings (EA- and ED-Series), AE-15 is a popular choice for transducers in the 0.5% accuracy class, intended for operation in the temperature range from -50° to +150° F *(-45° to +65° C)*. The main reason for this popularity is the relatively simple bonding procedure compared to that for the solvent-thinned epoxies described in the following section.

High-Performance Adhesives

Micro-Measurements offers three specially formulated epoxy (and epoxy-phenolic) adhesives for use in high-precision transducers — M-Bond 610, 600, and 43-B. These adhesives are designed for higher operating temperatures and thinner bondlines than the AE-10/15 system. They typically contain about 75% volatile solvents which must be liberated during specified air-dry periods and subsequent elevated-temperature curing cycles. The large solvent content allows the formation of extremely thin, hard and void-free bondlines, and greatly increases the pot life after mixing. The cement thicknesses under properly bonded gages are only about 0.0002 in *(0.005 mm)* with these adhesives. Obtaining such thin bondlines requires, however, strict adherence to recommended bonding procedures. Particularly important are the cure temperature (along with its rise time and duration), the clamping pressure, and the post-cure.[4] Micro-Measurements Instruction Bulletin B-130 should be consulted for the detailed step-by-step bonding procedure.

For transducer applications, M-Bond 610 epoxy-phenolic adhesive provides the highest level of performance, and is therefore the most widely used of the three solvent-thinned types. M-Bond 600 is similar in character and appearance to 610 except that it has been formulated for greater reactivity and correspondingly lower-temperature cure schedules. The increased reactivity naturally results in a shorter pot life (two weeks, compared to six weeks for M-Bond 610). The different formulation also contributes to greater variation in bondline thickness which, in turn, may have a detrimental effect on creep scatter in finished transducers. Because of this and other minor performance differences, the 610 adhesive is generally preferred over M-Bond 600 for transducer applications.

M-Bond 43-B is a single-component epoxy adhesive, developed particularly for compatibility with transducer strain gages. It is capable of thin, hard bondlines, much like M-Bond 610. Since it is a single-component adhesive, its pot life is the same as its shelf life, or about nine months at room temperature. M-Bond 43-B usually requires higher cure temperatures than either 610 or 600, and this may prevent its use on certain transducer alloys. Heat-treated aluminum alloys, for example, may have their properties altered by the relatively high-temperature curing and post-curing procedures specified for 43-B adhesive.

With carefully considered strain gage selection, guided by the criteria described in this and the preceding section; and employing a solvent-thinned high-performance epoxy adhesive; the user will have made two of the important (and requisite) steps toward achieving a stable, accurate transducer. Proper gage-installation technique is also a critical factor in transducer performance, of course; but the topic is not treated here, since thorough step-by-step procedures for gage bonding are given in the Micro-Measurements instruction bulletins accompanying each of the adhesives.

---

[4]An important point, often overlooked, is the significance of elevated temperature during the curing cycle. Single-component epoxies *always* require elevated-temperature cures; and many two-component epoxies also *must* be cured above room temperature for development of full properties. The latter types of adhesives often require that temperature elevation for the curing cycle begin not long after the end of the nominal, mixed, pot life. This is necessary to avoid inadequate adhesion and poor elevated-temperature properties in the cured state. M-Bond AE-15, 600, and 610 are examples of adhesives which require timely initiation of the elevated-temperature cure following application.

5

Completing The Basic Transducer

For reliably accurate measurements, careful consideration must be given to the remaining important steps in arriving at a completed load cell.

Previous chapters of this monograph have concentrated on the two primary components of a load cell — the spring element and the strain gage (and adhesive) system. To arrive, however, at a completed load cell which is capable of reliably accurate measurements, there are several other areas of transducer technology that require careful consideration and meticulous implementation. These include gage-circuit wiring and soldering; bridge output adjustment and compensation; and, finally, environmental protection of sensitive circuit elements. This chapter provides a compact treatment of all three areas.

Wiring

As already illustrated, modern strain gage transducers commonly employ four strain gage grids electrically connected to form a Wheatstone bridge measuring circuit. Depending on the transducer design, the same basic circuit could include 8, 16, or a still greater number of gages to achieve the desired result. This means that even the simplest full-bridge transducer circuit (Fig. 5.1) contains at least 12 electrical connections. The more complex designs might involve over 30 connections. Most of these connections, as shown in the figure, are inside the bridge circuit, where any small instability in resistance can severely affect transducer performance.

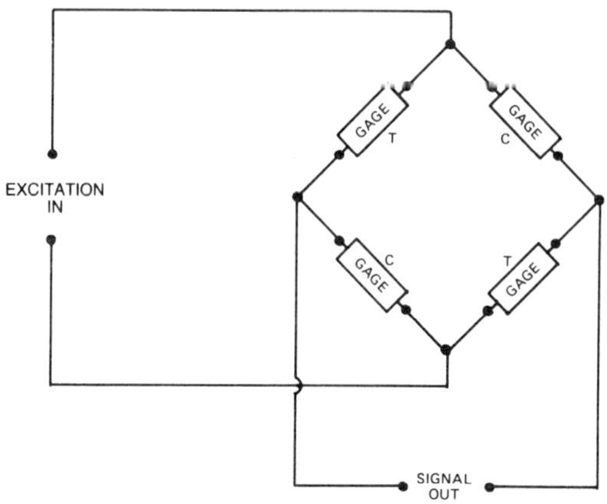

Fig. 5.1 — Basic full-bridge circuit typically used in strain gage transducers.

The usual method of connecting the gages and leadwires together into the bridge circuit is by soft soldering. Conventional procedures use a soldering pencil, solder, and a fluxing agent. Some work has been done experimentally with ultrasonic wire bonding and laser soldering, but the standard practice in the transducer manufacturing industry is to hand-solder the connections. A comprehensive treatment of soldering technology is beyond the scope of this publication, but the more important aspects of transducer wiring can be outlined as follows:

1. Items to be soldered must be *clean*, that is, free from oxidation and from adhesive or other films which could inhibit surface wetting by the molten solder.
2. Pure or activated rosin flux is the type normally recommended. Water-based acid fluxes may be required when solder connections are being made directly to the tabs of K-alloy (modified Karma) strain gages. In either case, the subsequent procedures for neutralizing and cleaning away the flux residues must be appropriate for the type of flux employed. ***NEVER USE*** the kind of paste flux (containing petroleum-base grease) that is widely sold in hardware stores for plumbing purposes. The residue from this type of flux can *never* be completely removed from strain gage installations.
3. Use a temperature-regulated soldering iron — set to a temperature no higher than necessary to make the solder melt and flow properly.
4. Tin the gage tabs and wire ends before soldering them together. In doing so,

the soldering pencil tip should be allowed to dwell on the surface no longer than required to flow the solder; maximum 1 to 2 seconds.

5. Join the wires to the gage tabs, using additional flux if necessary. Connections should form smooth, shiny pillows of solder, without spikes or round lumps. As in the tinning operation, the tip of the soldering pencil should dwell on the joint only long enough to flow the solder.
6. Intergage wiring should be symmetrically arranged, and of the same length from gage to gage. Wherever possible, the wires should be bonded to the spring element with a soft material such as M-Coat C. This will help keep the wires and gages all at the same temperature.
7. Use a wire size which is consistent with the solder tab size. A typical wire for intergage connections is #34 AWG *(0.16 mm)* solid copper, polyurethane-insulated. The wire size should be increased if the distance between gages is more than 3 to 4 in *(75 to 100 mm).* Stranded wire, because of its noise-generating tendency, should be avoided in the bridge circuit if at all possible.
8. Remove all traces of soldering flux. Residual flux is a source of contamination and possible corrosive attack. This topic is of first-rank importance, and is discussed additionally under *Environmental Protection.*

Following these procedures, a completely wired, dual bending-beam transducer would have the appearance shown in Fig. 5.2.

Circuit Refinement

The basic strain gage and bridge circuit arrangements of Figs. 5.1 and 5.2 are usually adequate for low-accuracy do-it-yourself transducers. They can also serve very satisfactorily in integrated transducer/instrument systems where the capability for automatic adjustment, correction, and scaling of the circuit output can readily be incorporated in the instrumentation (see page 57). This is not true in general, however, for a large class of precision commercial transducers. The

Fig. 5.2 — Representative spring element with gages installed and wired (prior to application of protective coating).

reference here is to transducers of the "stand-alone" type; e.g., off-the-shelf units intended for use with a wide variety of instrumentation systems. To achieve the required accuracy in such transducers, it is necessary to introduce auxiliary or supplementary resistors into the circuit as means for adjusting the circuit output and compensating for the effects of temperature changes.[5]

There are four principal aspects of transducer behavior that commonly require such "tuning":

1. Zero shift with temperature change.
2. Initial zero balance.
3. Span shift with temperature change.
4. Initial span adjustment (calibration).

Expanded descriptions of these topics are given in the following paragraphs.

Zero Shift with Temperature Change — It is well known that strain gages respond not only to strain, but also to temperature. Because there is a gage "output", or resistance change, with temperature, the effect is often referred to as temperature-induced apparent strain or, currently, *thermal output*. Changes of gage resistance with temperature result both from the temperature coefficient of resistance of the grid, and from the usual difference in thermal expansion coefficients between the grid and transducer spring element. With modern self-temperature-compensated (S-T-C) strain gages, the major portion of this effect has been suppressed, but it cannot be eliminated entirely over any appreciable temperature range.

In theory, however, when the gages are arranged in fully active half- or full-bridge circuits, as they normally are in transducers, the thermal outputs of corresponding gage pairs in adjacent bridge arms should exactly cancel by subtraction, since they should be the same in both magnitude and sign. Thus, if the bridge were initially balanced for zero output with no load on the transducer, it should remain so, irrespective of temperature change. Practically, however, it must be recognized that no two gages — even from the same manufacturing lot — are perfectly identical; nor, in general, are their installation conditions. As a result, temperature change in a transducer ordinarily causes a small bridge unbalance, called "zero shift with temperature".

Initial Bridge Balance — Because of normal gage-to-gage resistance variations, and additional resistance changes produced during gage installation, the bridge circuit is usually in a resistively unbalanced state when first connected.

[5]Dorsey, J., "Homegrown Strain-gage Transducers." Presented at 1976 SESA Spring Meeting. Available upon request from Measurements Group, Inc.

Bridge Output Adjustment and Compensation for Integrated Transducer/Instrument Systems

When the transducer forms an integral part of a measurement system having its own "dedicated" (i.e., task-specific) electronic package, some or all of the output adjustment and compensation operations can be performed in the instrumentation, instead of in the transducer circuitry as described in the main text. This type of system may include, for example, an analog-to-digital converter to digitize the transducer signal, and a microprocessor (sometimes with RAM) for the execution of control and logic functions. Thus, in the more sophisticated systems, all of the output correction features required by the transducer can easily be incorporated in the instrumentation; and the transducer circuit can be as simple as that shown in Fig. 5.1, with no auxiliary resistors needed.

Integrated transducer/instrument systems are widely used in weighing, parts-counting, and similar scale applications. Consider, for such cases, the problem of establishing the scale zero in the presence of a potentially variable or unpredictable tare due to the container weight. Under these conditions, after suppressing the effect of the tare weight, the zero indication on the scale obviously does not represent zero load on the transducer. Thus, there is nothing to be gained by initially adjusting the bridge circuit to achieve an accurate resistive null-balance for the unloaded transducer. Instead, the signal to the system display (or other output mode) is zeroed in the instrumentation — either automatically, or on command, varying with the nature and purpose of the system.

A similar situation exists for the compensation of zero shift with temperature. If the rate of temperature change in the transducer is low with respect to the frequency with which measurements are made, any zero shift which occurs in the bridge circuit will be eliminated each time re-zeroing is done by the instrumentation. When this is not the case, compensation can be achieved with the help of a temperature sensor installed on the spring element in the area of the strain gages. With an appropriate correction function built into the instrumentation, the output of the temperature sensor can be used to maintain continuous compensation for zero shift.

Compensation for span shift with temperature change can also be accomplished by installing a temperature sensor on the transducer spring element to provide the instrumentation with continuous temperature input. It is necessary in this case, of course, that a representative transducer assembly be calibrated for its span-shift characteristics to permit designing the proper compensating function into the instrumentation.

The initial span of the integrated system is readily set during manufacture by adjusting the overall sensitivity to produce a predetermined output signal level or indication corresponding to full scale load.

With stand-alone transducers, for purposes of general applicability with differing instrument systems, it is advantageous to have the bridge circuit initially balanced, or approximately so. A typical transducer of this type would commonly be balanced so that its electrical output at no load deviates from zero by no more than ±1% of full-scale output.

Span Shift with Temperature Change — The normalized output of a transducer (ratio of electrical output to unit mechanical input) is referred to as its *span*, and is usually expressed in mV/V output at full-scale load. Unless compensating measures are taken, the span ordinarily changes with temperature. The change in span occurs because both the gage factor of the strain gages and the modulus of elasticity of the spring element material are functions of temperature. Consider, for instance, constantan strain gages installed on a steel spring element. For an increase in temperature of, say, 100° F *(55° C)*, the gage factor rises by about 0.5%, and the modulus of elasticity of the steel decreases by about 1.5%. The result is an increase in transducer span by approximately 2.0%.

In the case of K-alloy gages, on the other hand, the gage factor decreases with a temperature rise, partially offsetting the effect of decreasing elastic modulus, and thus reducing the span shift. By special alloy selection and foil processing, this attribute is exploited in Micro-Measurements EMC (Effective Modulus Compensation) gages to largely eliminate span shift for specified spring element materials. Space limitations do not permit further description of EMC gages here, but additional information can be obtained by contacting the Measurements Group Transducer Applications Department.

Initial Span Adjustment — The nominal span of a transducer is, of course, one of the initially specified design parameters. In the production of the transducers, however, individual units of the design will exhibit small variations in span due to the tolerances on the mechanical and electrical components involved. With stand-alone commercial transducers, it is usually necessary that the span be set accurately to some predetermined value. To provide for this, the standard practice is to incorporate means for span adjustment in the circuitry associated with the bridge.

Auxiliary Resistors for Output Adjustment and Compensation

For a stand-alone transducer to operate accurately with different instrument systems, all of the foregoing adjustment and compensation functions must obviously be integral with the transducer proper. This is normally accomplished by permanently inserting several different types of special resistors at selected locations in the transducer circuitry. These ohmic components, carefully specified for resistance value and thermal coefficient of resistance (TCR), may be in any of several forms: wire (ranging from short lengths to bobbins or spools), bondable resistors (either adjustable or fixed-value types), or discrete resistors in small plastic or metal cases.

Wire resistive elements offer the advantages of low cost and wide resistance range, with the latter readily obtainable by adjusting the wire length and/or diameter. Bondable resistors provide the additional advantage of very close thermal tracking of transducer response, since they are in intimate contact with the spring element surface. Regardless of the resistor construction used, there are two basic types needed: temperature-sensitive resistors, and those that are relatively insensitive to temperature. These are often referred to as "high-TC" and "low-TC" resistors, respectively. Low-TC resistors are required for initial circuit adjustments to achieve zero bridge balance and to set the transducer span at the desired level. Compensations for zero shift and span shift with temperature are accomplished with high-TC resistors. The resistance alloys commonly employed for low-TCR applications are constantan, Karma, and manganin. In contrast, copper, pure nickel, and Balco are the usual choices for high-TCR needs.

For all adjustment and compensation procedures except, perhaps, span shift, the appropriate value of the resistor is determined experimentally; and this value typically varies from transducer to transducer. As a result, in order to expeditiously meet their production requirements, the major transducer manufacturers often stock many different resistor values in the types being used. With smaller manufacturing operations, and for prototype or one-of-a-kind applications, adjustable resistors offer the more practical means for refining the transducer characteristics.

Three types of bondable adjustable resistors are illustrated in Fig. 5.3, with pattern (configuration) designations **C**, **D**, and **E**. All three types are based on "ladder" arrangements which allow the user to easily adjust the resistance (upward) by cutting away some of the "rungs" or steps in the ladder. Using a

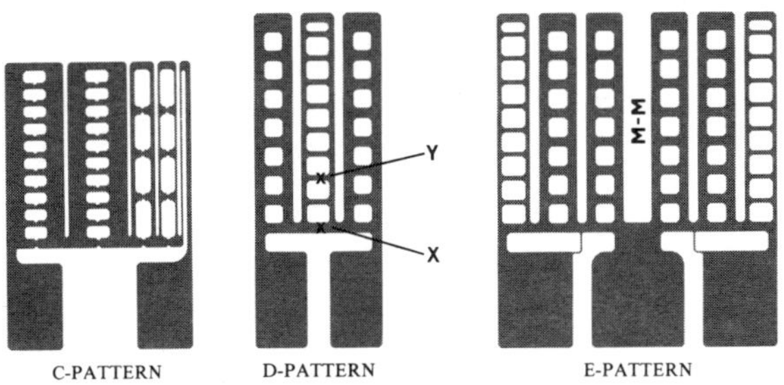

Fig. 5.3 — Bondable adjustable (ladder) resistor patterns used for bridge output adjustment and compensation.

scalpel or other sharp tool, the horizontal webs are severed and removed to lengthen the conducting path, and thus increase the resistance.

In the **D** pattern, for instance, cutting step *X* would increase the resistance by about 6% of the maximum obtainable value. Progressively cutting upward on this same ladder (leaving the top rung intact) would add about 6% of the maximum resistance for each step cut. The **D** pattern also has two outside ladders which, when cut progressively upward from the bottom, produce a finer adjustment equal to approximately 3% per cut. Selectively skipping steps of these ladders (i.e., cutting step *Y* after step *X*) produces finer adjustment in resistance. This feature gives the user a great deal of flexibility in gradually converging on the required final resistance. It can be seen from the figure that the **E** pattern is effectively similar to two **D** patterns placed side-by-side and joined with a common terminal. This configuration is designed for use at a corner of the bridge circuit, with one of the resistor halves in each of two adjacent bridge arms.

The **C** pattern in Fig. 5.3 has somewhat different characteristics than the **D** and **E** patterns, partly because of the fixed minimum resistance associated with the single vertical filament near the right-hand edge. It is also different in the geometry of the ladders. With this configuration, the right-most ladder provides four cutting steps, each of which represents about a 20% increase from the uncut initial resistance. Similarly, the second ladder has four 10% steps, and there are twenty 1% steps in the remaining two ladders. As for the **D** and **E** patterns, step removal from near the top of a ladder causes a smaller resistance increment than does removal from near the bottom. The **C** pattern can be used whenever the minimum required resistance will be about half of the fully cut maximum. When it is not known whether any resistance will be needed, the **D** pattern is the better choice because of its very low uncut resistance.

Bondable adjustable resistors are installed on the transducer spring element, wired, and protected from the environment using the same procedures as for strain gages. For optimum thermal tracking, the high-TCR compensation resistors should normally be located as close as possible to the strain gages, but preferably in a low strain field. Wiring which is within the bridge circuit should be symmetrical in length and placement, and should be in good thermal contact with the spring element.

How to Compensate Zero Shift With Temperature

Assume, to begin with, that four strain gages, which should always be of the same type and from the same manufacturing lot, are bonded to a spring element and wired to form a full-bridge circuit. When excitation is applied, the output of the bridge will not ordinarily be zero, but should be small enough to fall well within the working range of the indicating instruments typically used with transducers. Then, if the gaged spring element is warmed or cooled, and stabilized at some other temperature, the output will normally change. This characteristic behavior of an uncompensated transducer can be readily verified by experiment. Even in cases where the spring element is part of a large structure such as a bridge or a commercial aircraft, it is usually possible to perform such a test by taking measurements at different times in the day when the temperature changes naturally.

After carefully measuring the change in output with temperature, the nominal value of the compensating resistor can be calculated directly for any selected high-TC resistor alloy. Assume, for instance, that the bridge circuit is composed of four 350-ohm gages with a gage factor of 2.0, and that a temperature rise of 100° F *(55° C)* causes a change in bridge output equivalent to 35 μin/in *(μm/m)* in a single active gage. The corresponding resistance change can be expressed as:

$$\Delta R = GF \times \epsilon \times R_B \tag{5.1}$$

where: ΔR = resistance change, ohms

GF = gage factor of gages

ϵ = equivalent output strain

R_B = bridge resistance

Substituting the above quantities,

$$\Delta R \times 2 \times 35 \times 10^{-6} \times 350 = 0.0245 \text{ ohm [per } 100^\circ\text{F } (55^\circ\text{C})]$$

Copper, with a TCR of about 22% per 100° F *(55° C)* is a suitable material for the compensating resistor in this case. When the required compensating resistance change and the TCR of the resistor are both expressed in terms of the same temperature range, the copper resistance can be calculated directly from:

$$R = \frac{\Delta R}{TC} = \frac{0.0245}{0.22} = 0.11 \text{ ohm}$$

Thus, a copper resistor of 0.11 ohm in series with the gage in one of the arms should exactly compensate for the temperature-induced zero shift of the bridge circuit. The resistance of the copper required is usually very small, and this can make the task of selecting a resistor and soldering it in place quite difficult. In the present example, for instance, if AWG-40 *(0.08-mm)* copper wire were used for the resistor, the length to produce 0.11 ohm would be only about 1.25 in, or 32 mm.

Selecting, instead, a copper double ladder resistor (**E** pattern) for installation in one corner of the bridge circuit (Fig. 5.4a) offers several advantages. The initial uncut resistance of the pattern is very low; and the arrangement indicated in the figure assures that an adjustable resistor will be available in whichever bridge arm requires it. Furthermore, if over-compensation accidentally occurs, it is easy to correct the error by simply increasing the conductive path length of the copper in the adjacent arm.

Once the spring element is gaged and the bridge wired with the copper double ladder in place, the initial bridge output can be measured. Then the unit can be placed in an oven or environmental chamber to determine the output at a second temperature. The direction of the change in output (with respect to the temperature change) will indicate which of the two adjacent arms needs added temperature sensitivity. Through experience, and trial and error, the ladder in that arm can be increased in resistance until the temperature-induced zero shift is within acceptable limits.

How To Adjust For Initial Zero Balance

The procedure for adjusting the no-load bridge output to zero is essentially the same as for zero-shift compensation except that a low-TC resistor (e.g., manganin) is needed.

Assume, for an example, that after completing the zero-shift compensation, the bridge exhibits an initial unbalance equivalent to 800 μin/in *(μm/m)* in a single arm. Substituting into Eq. (5.1) again:

$$\Delta R = 2 \times 800 \times 10^{-6} \times 350 = 0.56 \text{ ohm}$$

Thus, the resistance of one bridge arm needs to be increased by 0.56 ohm to null the output. If manganin wire in AWG-37 *(0.08-mm)* size were used, this resistance would correspond to a length of about 0.45 in *(11.5 mm)*. Depending on the direction of the initial unbalance, the small wire can be inserted in series with the gage in one arm of the bridge for approximate zero balance.

An alternate method is to install a double ladder resistor such as the **E** pattern in the bridge corner opposite the copper ladder resistor. This arrangement, shown in Fig. 5.4b, permits easy, precise adjustment of the initial zero balance. If the strain gages in the bridge circuit are made from constantan, then the usual choice for the **E** pattern resistor is also constantan to minimize the introduction of added zero shift with temperature. Similarly, a K-alloy resistor is normally chosen for use with K-alloy strain gages.

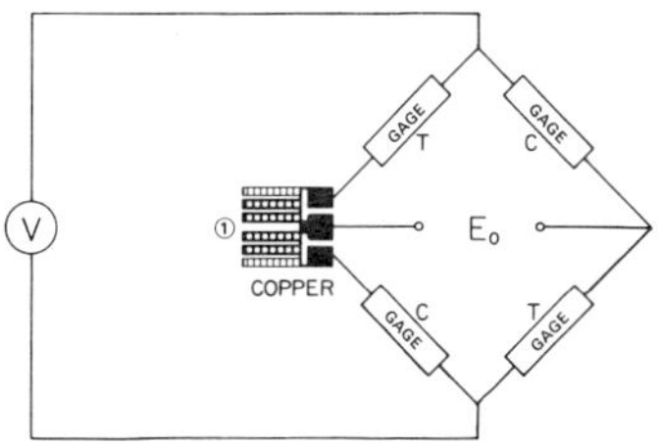

a. — High-TCR copper resistor (#1) inserted in corner of bridge circuit, and adjusted to maintain bridge balance over the operating temperature range.

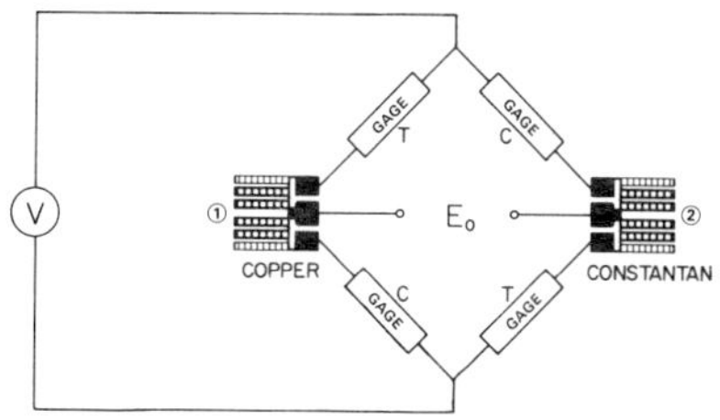

b. — Low-TCR constantan resistor (#2) inserted in second corner of bridge circuit, and adjusted for initial zero balance.

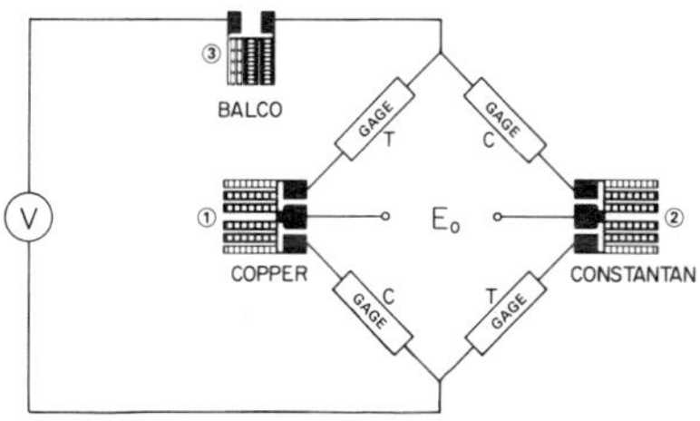

c. — High-TCR Balco resistor (#3) inserted in bridge power supply line, and adjusted to maintain essentially constant transducer sensitivity (span) over the operating temperature range.

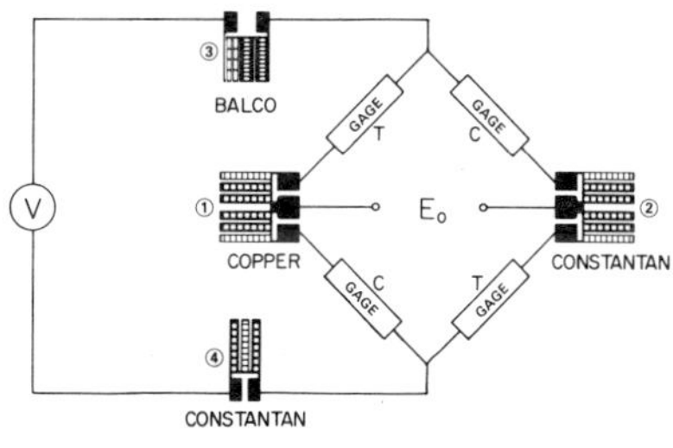

d. — Low-TCR constantan resistor (#4) inserted in bridge power supply line, and adjusted to set the initial span at the desired calibration level.

Fig. 5.4 — Use of auxiliary resistors for bridge output adjustment and compensation.

How To Compensate Span Shift With Temperature

Compensation for span shift with temperature can be accomplished by inserting a temperature-sensitive (high-TC) resistor in the bridge excitation supply line (Fig. 5.4c). When the resistor is of the proper size and temperature sensitivity, it serves to vary the bridge voltage so as to counteract the temperature-dependent changes in bridge output. With a temperature rise, for example, the bridge output for a particular load tends to rise also. This can be compensated for, however, if the resistance in the supply line increases simultaneously to produce an offsetting reduction in bridge voltage. The procedure for compensating span shift is usually the most difficult adjustment to make because it requires accurately loading the transducer and making measurements at two or more different temperatures.

Nickel resistors have been widely used for span shift compensation, but there has recently been considerable use of Balco, a very stable nickel-iron alloy with a relatively high temperature coefficient of resistance. Balco offers several advantages over pure nickel. It is less expensive, is easier to fabricate, and has about 2-1/2 times the resistivity of nickel. One of its few disadvantages is that it has a slightly lower temperature coefficient of resistance than nickel at typical transducer working temperatures (above –100° F, or –75° C). Given the TCR of the nickel or Balco resistor material, and the rate of change of transducer output with temperature (both expressed in the same units, such as percent per degree), the nominal value of the span-shift compensation resistor can be calculated with the following relationship:

$$R_M = \frac{\Delta S \times R_B}{\alpha_M - \Delta S} \tag{5.2}$$

where: R_M = resistance of span-shift compensation resistor

ΔS = percent change in span per unit temperature change

R_B = bridge resistance

α_M = temperature coefficient of resistance of compensating resistor (in percent)

For preliminary estimation of the required resistance, prior to running span tests, the change in span with temperature can be estimated from:

$$\Delta S = -(\Delta E) + \Delta GF \tag{5.3}$$

where: ΔE = percent change in elastic modulus of the spring material per unit temperature change (normally negative)

ΔGF = percent change in gage factor per unit temperature change

Table 5.1 gives nominal values of ΔE, ΔGF, and α_M which can be used with Eqs. 5.2 and 5.3 for preliminary calculation of the compensating resistor. The same equations have been incorporated in Fig. 5.5 (for Balco) to permit directly reading the required resistance from the graph — given the type of gage in use, and the modulus coefficient of the spring material. To find the equivalent resistance in nickel, it is only necessary to multiply the Balco resistance by $24/33 \approx 0.7$ (the approximate ratio of their respective TCR's). As a numerical example, consider a spring element made from aluminum alloy, with four 350-ohm constantan gages installed. Referring to Fig. 5.5, the nominal value of a Balco compensating resistor is 60 ohms. If a nickel resistor is to be used, it should have a resistance of 0.7 x 60 = 42 ohms.

Table 5.1 — Material Properties

ΔE	
spring element	temperature coefficient of modulus
17-4 PH stainless steel	-0.015%/°F *(-0.027%/°C)*
tool steel	-0.013%/°F *(-0.023%/°C)*
beryllium-copper	-0.025%/°F *(-0.045%/°C)*
aluminum	-0.030%/°F *(-0.054%/°C)*
ΔGF	
gage alloy	temperature coefficient of gage factor
constantan	+0.0050%/°F *(+0.0090%/°C)*
K-alloy (06 comp.)	-0.0057%/°F *(-0.0103%/°C)*
K-alloy (13 comp.)	-0.0083%/°F *(-0.149%/°C)*
α_M	
resistor material	temperature coefficient of resistance
nickel	+0.33%/°F *(+0.59%/°C)*
Balco	+0.24%/°F *(+0.43%/°C)*
copper	+0.20%/°F *(+0.36%/°C)*

The preceding method will not generally result in precise compensation for span shift because of the approximate properties of the spring element, gages, and resistor alloys used in the calculations, and in Fig. 5.5. Furthermore, re-adjustment of the compensation resistor will usually be required when the span-set or calibration resistor is inserted as described in the following section. The most practical approach is to determine the nominal resistance from Fig. 5.5, and then select a **C** pattern resistor which has an initial, uncut resistance below the required value, and a fully cut resistance above it. By successive span-versus-temperature tests, the resistor can be adjusted to minimize the observed span changes. It is usually possible, with this procedure, to adjust span shift versus temperature to less than 0.001% of signal per °F *(0.0018%/°C)*.

How to Adjust Span

The final circuit adjustment is to set the span to some predetermined level. This is accomplished by inserting a temperature-insensitive (low-TC) resistor in series with the bridge excitation supply (as shown in Fig. 5.4d). A suitable resistor configuration for this purpose is the *D* pattern, with the alloy (constantan or K-alloy) selected to match the strain gage alloy. The nominal resistance can be determined by measuring the total impedance of the circuit in Fig. 5.4c (including the Balco span-shift resistor) and then calculating the further attenuation needed to bring the bridge output to the desired span. With the span-set resistor (#4 in Fig. 5.4d) in place, successive span tests and resistor adjustments can be used to converge on the specified output level.

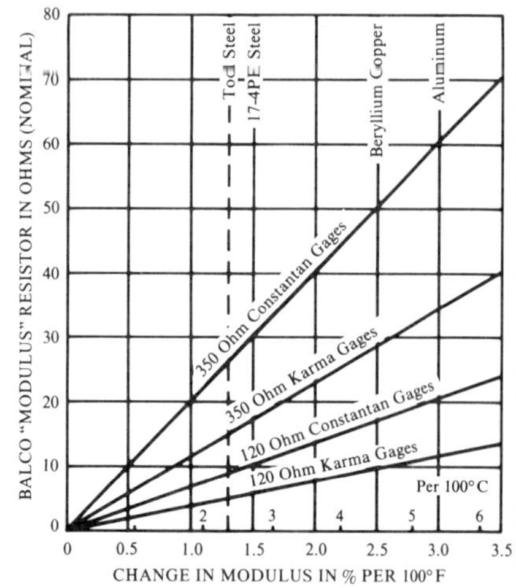

Fig. 5.5 — Calculated Balco resistance for span-shift compensation.

In actual practice, the most expedient procedure for accomplishing the foregoing circuit refinements is to initially calculate the approximate values of the span-shift compensation and span-set resistors (#3 and #4 in Fig. 5.4d), and then install and wire all four resistors on the spring element along with the strain gages. Subsequently, and prior to applying a protective coating, the resistors can be adjusted as described here to obtain the desired transducer performance.

It should be realized that the foregoing procedures for output adjustment and compensation account primarily for first-order effects on transducer performance. The achievement of still greater precision necessitates further circuit refinement, usually involving the use of more auxiliary resistors. In compensating span shift with temperature, for example, the inserted high-TC resistor normally introduces a degree of nonlinearity in the output. Reducing the latter effect requires, in turn, additional linearizing resistors. Compensation techniques for higher-order effects such as this are beyond the scope of the present treatment.

Environmental Protection

Sensitive strain gage circuits require protection against most transducer environments. Even under indoor conditions, airborne moisture and low-level contaminants can adversely affect long-term stability. For use in less sheltered applications, many commercial transducers must be certified to withstand intermittent water splash or short-term immersion with no degradation in accuracy. Selection of the most suitable environmental protection system for a particular transducer design requires careful consideration of several factors. The effectiveness of the system in providing a barrier between the circuit components and the intrusive elements in the environment is naturally a primary consideration. It is also critical that the protective system not affect transducer performance (linearity, stability, etc.), at least in any noncompensable way. Beyond these, there are always such practical considerations as cost and manufacturability.

In terms of barrier properties, the hermetic seal is obviously the ultimate in effectiveness. Organic protective coatings, applied over the gage circuits, are much less expensive, and therefore widely used in high-volume, lower-cost transducers. In either case, however, the barrier may be rendered useless (or worse) if the gage installation is not perfectly clean before sealing. Residual contaminants, left on the gages or compensation/adjustment resistors from circuit installation and soldering operations (or from subsequent exposure and/or handling), become permanently entrapped there. The result can be a slow, irreversible chemical attack on the very thin foils from which the components are made, leading to progressive loss of stability and degradation in transducer performance.

The most vulnerable gage installations to damage by contaminants are those with open-faced (bare, or unencapsulated) grids, since the gage grid is the primary area of sensitivity to slow corrosive attack. Encapsulated gages greatly reduce the probability of grid contamination, but their use does not obviate the need for thorough cleaning of the gage installation prior to sealing from the environment. When open-faced gages represent the best choice for a particular application, however, the cleaning operation should receive extra care and attention.

As indicated earlier, the gage installation should always be cleaned directly after completing the soldering operations. When using any kind of acid flux, it is particularly important that the flux residues be promptly cleaned away, and the surface neutralized. Even when "noncorrosive" rosin flux is used for soldering, the residue, consisting of excess flux and various reaction products, must be completely removed to avoid leaving ionic contaminants on the gage installation. In either case, the cleaning procedure must be tailored to the type of flux in use, since the chemicals in the fluxes are radically different, as are the by-products formed at the soldering temperature.

When an acid flux has been used, the installation can be washed with a liberally applied mild (2%) hydrochloric acid solution to dissolve the complex salts formed during soldering. After washing away the acid, the installation is rinsed with a water-based neutralizer. In the case of rosin flux, the installation should first be thoroughly cleaned with rosin solvent. But the use of rosin solvent alone may still leave a film of rosin binder on the installation. For the most thorough cleaning, the rosin solvent should be followed by a wash with a water-based semisolvent (e.g., denatured ethyl alcohol and water solution), and then rinsed with a water-based neutralizer.

Irrespective of the initial cleaning which may be done following the gage installation, it is absolutely essential that the ultimate cleaning operation be performed just before sealing the installation from the environment, whether by hermetic seal or protective coating. In both instances, after the application of water-based cleaning compounds, the installation should be thoroughly dried by warming the transducer prior to the sealing operation. It is particularly critical when using a protective coating that the transducer be both warm and dry when the organic compound is applied.

Hermetic Seals — The hermetic seal, when its cost can be justified, offers the only absolute (impenetrable) barrier between the gage installation and the external environment. Figure 5.6 illustrates two common arrangements for completely sealing the strain gage area within metal enclosures. Both configurations are designed for minimal mechanical restraint with respect to normal spring element deformations. Welding of the metal seals is the most effective method of closure, and permits the cavity to be subsequently purged with an inert atmosphere such as dry nitrogen or argon.

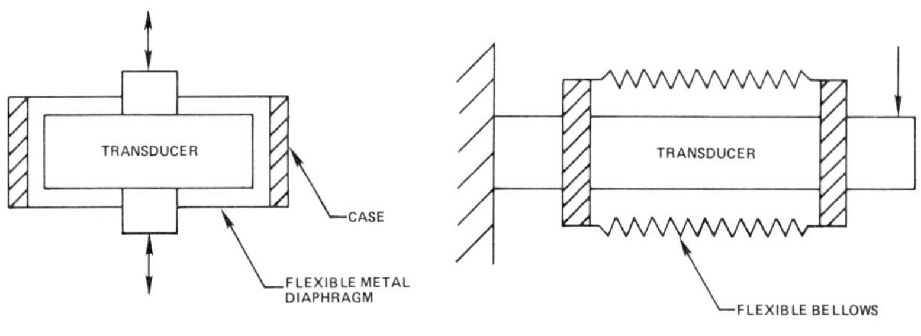

Fig. 5.6 — Typical hermetic seal designs for environmental protection of transducer gage installations.

The enclosed volume is sometimes filled with transformer oil or a silicone oil, but neither of these contributes to gage protection if the unit is, in fact, hermetically sealed. The principal functions of the oil in such cases are to improve thermal equilibrium in the spring element and to provide damping. When oils are used, they must naturally be *clean* (free from contaminants) and *dry*. Silicone oils, for example, can contain significant amounts of dissolved water which would be detrimental to the gages and to transducer stability. It is also necessary, with oil-filled cavities, to provide for expansion of the fluid with a temperature rise, and for displacement of the fluid due to deformation of the spring element. Simple immersion of the gaged section of the spring element in a free-surface oil bath offers little or no long-term protection from the environment. The practice is especially poor for silicone oils, since moisture can readily migrate through the oil to the gages.

Rubber diaphragms and bellows assemblies, although not totally impervious to air and moisture, can sometimes be used in place of metal. They are much lower in cost, and may serve adequately when environmental conditions are not too extreme. Attachment can be made through mechanical clamps or by adhesive bonding.

Wax — Microcrystalline wax, when conditions permit its use, is the most effective organic barrier material currently available for protecting strain gage circuits from moisture. Its principal disadvantages are its tendency to become brittle and crack at low temperatures, and its limited upper temperature range due to melting at 170° F *(80°* C). It is also very low in mechanical strength.

When using microcrystalline wax for transducer gage protection, the following precautions need to be observed:

1. The transducer must be warmed to 120° F, or 50° C (and, of course, be completely dry) when applying the wax.
2. The stiffness of the spring element at the gaged cross section should be great enough so that the transducer response is essentially unaffected by the presence of the wax coating.
3. All leadwires must be firmly anchored to the spring element before applying the wax, since the coating is easily cracked, and the seal broken, by leadwire motion or vibration.
4. The transducer casing or some other design feature must provide mechanical protection for the wax coating.

Butyl Rubber — Butyl rubbers and solvent-thinned butyl rubbers are widely used as coating materials for transducer gage protection. They provide a good moisture barrier (although inferior to wax) and, because of their characteristically low elastic modulus, do not reinforce the spring element excessively. In their natural form (small sheets), they have the consistency and stickiness of freshly used chewing gum. The softness and conformability of the sheets makes them easy to apply to the gage and surrounding area. Solvent-thinned butyl rubbers are supplied in liquid form for application by brush, and require a drying cycle before the transducer can be placed in service.

Silicone Rubber — For many transducer applications where prolonged contact with moisture is not a problem, RTV silicone rubber can serve adequately as the protective coating over a gage installation. Although it is the least effective moisture barrier among the materials described here, it has better mechanical strength and resilience than wax, and is easier to apply. An RTV primer is often used to improve coating adhesion to the spring element.

It should be noted that some silicone rubbers liberate acetic acid when curing. Since the acid may be harmful to the gage foils, such silicone compounds should be avoided in favor of the noncorrosive humidity-reactive curing types. The silicone rubbers with acetic acid can be identified readily by their strong, pungent, vinegar-like odor in the uncured state.

6

Beyond The Basics

Additional expertise is normally required to produce a high-quality precision transducer.

The material contained in this book represents what might be considered a comprehensive "primer" on strain gage transducers. It includes a sizable collection of spring element designs; background on spring element materials, and the criteria for material selection; strain gage and adhesive systems for transducer applications, with selection considerations; gage circuit wiring; bridge output adjustment and compensation; and, finally, environmental protection for transducers. The emphasis throughout, for purposes of coherence and continuity, has been on load cells, since most of the basic requirements for a properly functioning load cell are applicable in general to other types of strain gage transducers as well.

While the information presented here should provide a solid background for approaching the task of transducer design, additional expertise is normally required to produce a high-quality precision transducer. To begin with, there are many other, more subtle, considerations which influence transducer performance, but are outside the scope of the present work. Unfortunately, much of the know-how needed to refine a transducer design exists in the form of "art"; that is, unpublished knowledge and insight gained through long experience in transducer engineering. Such considerations can become particularly important when attempting to arrive at the optimum transducer for a given application — whether in terms of accuracy, resolution, cost, reliability, or other specifications which must be satisfied.

When difficulties are encountered in achieving the desired transducer performance, the problems can usually be resolved by consultation with the Measurements Group Transducer Applications Department. The same source can be called upon for information about the most recent advances in strain gage technology applicable to the continually evolving transducer field.

Index